TOTAL WATER MANAGEMENT
PRACTICES FOR A SUSTAINABLE FUTURE

TOTAL WATER MANAGEMENT

PRACTICES FOR A SUSTAINABLE FUTURE

Neil S. Grigg, PhD, PE

American Water Works Association

Total Water Management: Practices for a Sustainable Future
Copyright ©2008 American Water Works Association

All rights reserved. No part of this publication may be reproduced or transmitted in any form or by any means, electronic or mechanical, including photocopy, recording, or any information or retrieval system, except in the form of brief excerpts or quotations for review purposes, without the written permission of the publisher.

Disclaimer
This book is provided for informational purposes only, with the understanding that the publisher, editors, and authors are not thereby engaged in rendering engineering or other professional services. The authors, editors, and publisher make no claim as to the accuracy of the book's contents, or their applicability to any particular circumstance. The editors, authors, and publisher accept no liability to any person for the information or advice provided in this book or for loss or damages incurred by any person as a result of reliance on its contents. The reader is urged to consult with an appropriate licensed professional before taking any action or making any interpretation that is within the realm of a licensed professional practice.

AWWA Publications Manager: Gay Porter De Nileon
Technical Editor/Project Manager: Martha Ripley Gray
Cover Design/Production Editor: Cheryl Armstrong
Cover Photo: Colorado River Aqueduct outside of Los Angeles, ©iofoto @ Shutterstock

Library of Congress Cataloging-in-Publication Data

Grigg, Neil S.
 Total water management : practices for a sustainable future / by Neil S. Grigg.
   p. cm.
 ISBN-10: 1-58321-550-6
 ISBN-13: 978-1-58321-550-0
 1. Water resources development. 2. Water-supply--Management. 3. Sustainable development. 4. Environmental policy. I. American Water Works Association. II. Title.

 HD1691.G748 2008
 333.91--dc22
                              2008010834

American Water Works Association

6666 West Quincy Avenue
Denver, CO 80235-3098
303.794.7711
www.awwa.org

CONTENTS

LIST OF FIGURES   vii

LIST OF TABLES   ix

FOREWORD   xi

CHAPTER 1

TOTAL WATER MANAGEMENT: FROM VISION TO EXECUTION   1

What is TWM, really?	2
Why is TWM needed?	3
TWM is about leadership	5
Is there an environmental crisis?	7
Barriers to sustainability	9
The nature of TWM	10
TWM—more political than technical	12
Use of case studies to explain TWM	13
What does the book contribute?	14
How do utilities take the lead?	16
Summary points	17
Review questions	17
References	17

CHAPTER 2

WATER MANAGEMENT AND ITS IMPACTS   19

The water supply problem	20
The water quality problem	21
The environmental problem	22
The Tragedy of the Commons	23
Arenas for action of TWM	24
Why sustainability is a shared responsibility	26
Threats to sustainability	28
Players and the water management actions they control	29

How the players create impacts on water systems	30
Summary points	39
Review questions	40
References	40
Case Study—Maintaining Supply While Preserving the Resource	43

CHAPTER 3

TOTAL WATER MANAGEMENT: VISION, PRINCIPLES, AND EXAMPLES 55

Fundamental concepts and definitions of TWM	55
Beyond the definition: putting TWM to work	65
Principles of TWM	65
Summary points	89
Review questions	90
References	90

CHAPTER 4

PLANNING AND SHARED GOVERNANCE 93

Planning and shared governance in TWM	95
Water resources planning	96
Governance and shared governance	103
Defining roles and relationships	106
Integration and coordination through shared governance	106
Regionalization: its promises and challenges	109
Toward the future	110
Summary points	111
Review questions	112
References	112

CHAPTER 5

TRIPLE BOTTOM LINE REPORTING FOR WATER AGENCIES 115

TBL as sustainability reporting	117
TBL as multicriteria scorekeeping	118

Use of indicators in TBL reports	121
TBL reporting for water management	124
Status of TBL reporting in the water industry	127
Utility TBL reports	128
TBL results in a region	128
Compiling a TBL scorecard: the Sydney Water example	131
A US example: Seattle Public Utilities	132
Integrity in reporting	133
Summary points	133
Review questions	134
References	134

CHAPTER 6

VALUE AND COST OF WATER 137

How society balances the allocation of water resources	141
What is meant by the value of water	144
Societal versus individual decisions: the accounting stance	146
How society computes benefits and costs	148
Use of cost-effectiveness analysis	155
Balancing the uses	155
Why people do not recognize the value of water and what can be done	155
Summary points	157
Review questions	158
References	159

CHAPTER 7

ENVIRONMENTAL WATER: ASSESSMENT, VALUE, AND SUSTAINABILITY 161

Sustainable development and natural systems	162
State of the environment	163
Summary of environmental issues	168
What are the water needs of natural systems?	171
Water needs of natural system elements	172

Water management actions and impacts	177
Balancing environmental benefits and costs in TWM	179
Environmental monitoring and assessment	181
Assessment at the watershed level	182
Summary points	186
Review questions	188
References	188

CHAPTER 8

SOCIAL IMPACTS OF WATER MANAGEMENT 191

Classification of social impacts	193
A system for social indicators of water projects	194
Public health and safety	196
Equal opportunity	201
Community goodwill	201
Social impact analysis: an assessment tool	202
Social rights and social responsibilities	203
Summary points	203
Review questions	204
References	205

CHAPTER 9

LAWS AND REGULATIONS OF WATER MANAGEMENT 207

Law coordinates and regulates water management	208
How law determines management choices	210
Water law	212
Water laws by levels of government	214
Regulation in the water industry	228
Roles of courts	230
International water laws	230
Summary points	230
Review questions	232
References	232

CHAPTER 10

POLITICAL AND INSTITUTIONAL OBSTACLES TO TWM   235

An explanation of institutional factors	236
Examples of institutional problems	237
Water institutions	238
Examples of institutional obstacles to TWM	240
Water supply industry constraints	242
Nonpoint source pollution	243
A method for institutional analysis	245
Political model of water planning	246
Gap analysis and remedies	248
Roles and responsibilities	248
Summary points	250
Review questions	251
References	252

CHAPTER 11

ENVIRONMENTAL STEWARDSHIP, ETHICS, AND EDUCATION   253

About stewardship	254
Environmental ethics	256
Environmental education	256
Roles and responsibilities	258
Environmental leadership	260
Requirements for environmental education and ethics	260
Summary points	262
Review questions	262
References	263

CHAPTER 12

WATER INDUSTRY PROSPECTS AND POLICIES   265

Threats to the water industry	265
Where the water industry is heading	266
State of the practice of TWM	268
Roles and responsibilities	269

Institutional arrangements	270
Final word	272
Summary points	273
Review questions	274
Reference	274

APPENDIX A

AWWA AND AWWARF STATEMENTS ABOUT TOTAL WATER MANAGEMENT AND RELATED CONCEPTS   275

AWWA Policy Statement on Developing and Managing Water Resources	275
AWWA White Paper on Total Water Management	277
AwwaRF definition of Total Water Management (1996)	281
AWWA definition of Total Water Management, from the *Drinking Water Dictionary* (2000)	281
References	282

LIST OF ACRONYMS   283

INDEX   287

LIST OF FIGURES

1-1.	TWM as a balancing act	2
1-2.	Financial and outreach responsibilities of business and utilities	6
1-3.	Balance point for sustainable development	7
1-4.	The balance in water management	8
1-5.	Total water management: a systemic concept	11
1-6.	How TWM works	16
2-1.	TWM looking inward and outward	24
2-2.	Water industry and impact sources	25
2-3.	How TWM relates to large and small actions	27
2-4.	Dual risks of water utilities	28
2-5.	Water and related land management activities	31
2-6.	Sources and impacts of TWM actions	33
2-7.	Water cycle uses, discharges, and effects	38
2A-1.	Vienne River and estuary	44
3-1.	How a framework organizes principles and processes	57
3-2.	IWRM policy sectors and purposes	59
3-3.	Development of TWM/IWRM concepts	59
3-4.	TWM processes and principles	65
3-5.	Water manager's dilemma	68
3-6.	Coordination and balancing in TWM	76
3-7.	Coordination and allocation of water	78
3-8.	Assessment and reporting to counter Tragedy of the Commons	85
4-1.	TWM process with shared planning	95
4-2.	Planning by levels	95
4-3.	Features of the Water Resources Planning Act	98
4-4.	Phases of the water resources planning act	99
4-5.	Rational planning in a political environment	102
4-6.	Circles of responsibility for water planning	105
4-7.	Roundtable	110
5-1.	Business reports versus TBL reports	116
5-2.	TBL and the Balanced Scorecard	120
5-3.	Reporting by public companies and government utilities	122
5-4.	Packing information to create an environmental indicator	123
6-1.	A fictitious water market and auction	139
6-2.	How valuation of water differs by level	142
6-3.	How economics and finance differ	144
6-4.	Concepts of value	146
6-5.	Water management is a balancing act	148

7-1.	World population with projection to 2050	162
7-2.	Watershed showing natural and human systems	164
7-3.	TWM as a comprehensive approach to water management	165
7-4.	How assessment and decision-making relate to each other	181
7-5.	Micro and macro aspects of water planning	185
8-1.	Water industry and outreach to society	192
8-2.	Water and the hierarchy of human needs	194
8-3.	Some possible health effects related to water	200
9-1.	How water-related law has grown	208
9-2.	How law and regulations affect water management	211
10-1.	A few "institutional factors"	236
10-2.	Water resources decision process	247
12-1.	Water industry triangle	267

LIST OF TABLES

1-1.	The TWM framework	3
2-1.	WQ 2000 sources of water contamination	22
2-2.	Management actions that impact water resources	26
2-3.	Threats to natural water systems	29
2-4.	People and organizations impacting water resources	30
2-5.	Main players in water management	30
2-6.	Estimated water use in the United States in the year 2000 in million gallons per day (*mgd*)	34
2-7.	USEPA findings on water quality	36
2-8.	Main pollutants	36
3-1.	Two definitions of TWM	56
3-2.	EU Water Framework Directive compared to the US framework	63
3-3.	TWM processes, principles, and practices	64
4-1.	Attributes of a water resources planning process	96
4-2.	Beneficial and adverse effects of water resources development by category	104
4-3.	Players in the planning processes	107
5-1.	Scoring strategies by goal	118
5-2.	Preferable attributes of a system of indicators	122
5-3.	TWM goals and specific measures	126
5-4.	TBL achievements by issue	128
5-5.	TWM elements and possible indicators	129
5-6.	Seattle Public Utilities statements and TBL/TWM	133
6-1.	Issues as perceived by society and organizations	147
7-1.	Conclusions of the 2003 USEPA report	168
7-2.	Impacts on water quality as defined in Water Quality 2000	170
7-3.	Targeted watersheds and their features	186
8-1.	TWM contributions to social systems	197
8-2.	USEPA list of contaminants and their potential effects	199
9-1.	Responsibility by element of the TWM definition	209
9-2.	Examples of legal categories including water law	212
9-3.	The legal matrix by level	212
9-4.	Legal frameworks for water management tasks	214
9-5.	Programs of the Clean Water Act	216
10-1.	Institutional factors relating to water management	237
10-2.	Classification of incentives in the water sector	240
10-3.	Discussion of problems confronting TWM	241
10-4.	Gaps in need and strategies for fixing them	249
10-5.	TWM roles and responsibilities	250

11-1.	Knowledge and value requirements of TWM elements	261
12-1.	Assessment scorecard by TWM element	269
12-2.	Roles and TWM tools by sector	271

Foreword

The concept of Total Water Management (TWM) was developed in the 1990s because leaders in the water industry sensed that single-purpose water management was outdated and too limited. While utilities competed hard for water and the government developed more regulations, water supplies were still getting scarcer and more degraded.

The origins of the TWM concept stem back to AWWA policies on water resources management and were expressed in a 1994 white paper that was drafted by the Total Water Management Task Force. This task force represented several committees and reported to the Technical and Educational Council and the Water Utility Council. Its chair was Gerald S. Allen, who at that time was with Avatar Utilities in Coral Gables, Florida. At that time, AWWA white papers were written at the direction of the board of directors to address a topic quickly.[1]

When the TWM concept was developed, AWWA was reexamining its roles on several fronts. A principal idea behind TWM is that the water supply industry should take leadership in resource conservation and in considering water management across the entire hydrologic cycle. While it is difficult to resurrect historical events, evidence shows that debates within AWWA leadership circles led to TWM becoming a value for AWWA to promote, along with Safe Water for All People and Customer Satisfaction.[2] These three slogans appeared on the cover of AWWA *MainStream* in the mid-1990s.

The promise of TWM is that we can promote sustainable development by working together to manage water on the basis of natural systems within watersheds. TWM might seem like a nice idea without much practical use, or it can be a powerful tool to forge cooperation and create win-win solutions among water managers. Its emphasis on stewardship shows that all citizens must participate. Otherwise, the relentless impacts of land development will degrade water supplies in spite of the efforts of water managers.

TWM doesn't lend itself very well to presentation in a manual of practice or a handbook. It is not really a set of steps so much as it is a set of principles. This book organizes its ideas and concepts and can be

---

1 Thanks to Linda Moody, AWWA's volunteer and technical support coordinator, for this information. She also reported that white papers are rarely written now because hot topics can be addressed in other ways.

2 Thanks to Bob Wubbena (president of AWWA, 1995–96) for insight into AWWA policy debates of the early 1990s that resulted in new directions and water industry leadership in TWM and related areas.

used to create other presentation mechanisms, such as training materials, PowerPoints, and video presentations.

TWM offers many examples of ways to be involved. People using its principles might be utility engineers, operators, or administrators. They might also be planners, scientists, or other support staff. They might be running advocacy organizations that reach out to private citizens. They might be leaders in convincing citizens to improve water use and management by conservation or nonpoint source control. They could be involved in the fishing and recreation industries and be working to keep the waters pure and safe. They might be regulators looking for better ways to regulate. They could be in the water industry's large support sector and want tips about providing products or services.

In presenting the book, I'm reminded of the strands of history that make up the TWM ideas. I think of AWWA leader Abel Wolman, who started his engineering practice early in the twentieth century and who advocated collaboration and effective public administration until his death in 1989. I think of Gilbert White, Maynard Hufschmidt, and Ted Schad, water leaders who started in the 1930s and were involved in Roosevelt's New Deal, the 1960s Senate Select Committee on Water Resources, or the National Water Commission. I think of AWWA leader Dan Okun, longtime professor at the University of North Carolina, who trained many students for global water leadership. And there are many water leaders today, such as Bob Wubbena and others, who developed the concepts of TWM and who have visions of how to use new approaches to solve old problems and create a better and more sustainable future. To all these leaders this book is dedicated.

Finally, I want to acknowledge the great help received from AWWA's publishing staff. Colin Murcray was the publications manager when the book was initiated. Gay Porter De Nileon became publications manager while the book was in progress. Not only has she managed the production process efficiently; she has contributed many substantive ideas for the book and improved it materially. She was ably assisted by Martha Ripley Gray, who went through the whole book and made many improvements in substance, style, and grammar. I also thank Cheryl Armstrong, who did the typesetting and created a great cover for the book. Thanks to all of you.

Neil S. Grigg
Fort Collins, Colorado
April 16, 2008

CHAPTER 1

TOTAL WATER MANAGEMENT: FROM VISION TO EXECUTION

The media reports frequent stories of global climate change, pollution, flooding, and the suffering they cause. Meanwhile, a quiet revolution is being led by men and women who care about sustainable use of water resources, public service, and a healthy balance between business and government. Many of these water leaders work in water supply and wastewater utilities or water management agencies.

Balancing water management and the environment is not only essential to a sustainable future, it's also good business. As Sandra Postel (2007) explained: "As one of the most publicly visible stewards of the earth's water sources, drinking water utilities are uniquely positioned to exert a leadership role in the emerging field of ecologically sustainable water management. In important ways, this field is integrating the traditional goals of water management with those of ecosystem conservation in order to sustain a broader spectrum of the valuable goods and services on which human communities depend."

The revolution is directed toward new ways to manage water resources and the public's business. That part of it addressed by this book is Total Water Management, or TWM. TWM offers to water utility managers and others involved in the water industry powerful and urgently needed tools to balance needs of water management and the environment.

Total Water Management means stewardship and management of water on a sustainable use basis. Its concepts are explained in detail in

2 TOTAL WATER MANAGEMENT

Figure 1-1.   TWM as a balancing act

chapter 3. TWM challenges water managers to juggle objectives that may conflict with each other (Figure 1-1).

What is TWM, really?

TWM is not a new and secret weapon. It is a new way of using tried-and-true methods to create a framework for principles and practices of sustainable water resources management. In explaining it, a working group of water utility officials defined TWM as the "exercise of stewardship of water resources for the greatest good of society and the environment" (AwwaRF, 1996).

A framework is a basic arrangement of a set of elements. It is a structure on which to hang the elements that make up the whole of your construct, which in this case is a method to manage water called TWM. For example, the European Union uses frameworks to construct bodies of law and policy to govern sectors of society and the economy. In the case of water, it is called the Water Framework Directive, and how it works will be described in more detail in chapters 3 and 9.

The TWM framework, outlined in Table 1-1, has a number of elements and good practices for stakeholders and participants in the water management game.

Table 1-1. The TWM framework

Participants	TWM provides
Utilities and water service providers	Guidelines to balance water supplies and impacts on the environment. TWM does not focus on business processes but it supports them.
River basin and watershed planners	Ways to cooperate and work together. It advocates management on a natural systems basis for watershed planning.
Regulators	Ways to blend regulatory strategies with volunteer actions to achieve higher levels of compliance than with command and control alone.
Government and policy community	Consistent ways to structure policy and government actions to support effective and shared governance.
Water users and people impacting water	Ways to integrate control of nonpoint sources and hydromodification with water storage, diversion, and point source discharges.

Why is TWM needed?

TWM is needed because the capacity of the environment to bear its load may be nearing its limit, and we cannot afford to waste or misuse water. Whether the topic is global warming, rising water demand, or exotic pollution, people sense that we must lighten the load and use sustainable management.

Unfortunately, the real world places barriers to doing this. In a perfect world, we could apply new technologies and create a society that places smaller burdens on natural systems. That's the goal of sustainable development, which is the concept of using resources wisely to preserve them for the future. But can the visionary concept of sustainable development be translated into action? The jury is out on the question, but whatever the outcome, the water industry will have a big part in it.

Leaders in the water industry have big, big roles to play in sustainable development. The playing field is changing rapidly. Students who are preparing right now to lead the industry will face a different set of challenges than the baby boomers did. The grandparents of baby boomers saw the close of the nineteenth century, before most piped water supply was available and when life expectancies were lower due to waterborne diseases such as cholera and typhoid. Baby boomers entered a stable workforce. Not so with today's new ball game.

Yes, the water industry faced many challenges in the twentieth century, and it witnessed dramatic improvements in water management and public health. Water safety improved, new methods to divert water and create supplies were developed, and many new laws and regulations were passed. But every advance was met with another challenge. Chemical contaminants increased, new forms of pollution have been identified, and impact on the environment from water use has increased, and neither government nor the free market seems able by itself to steer the water industry toward sustainable development.

Sustainable development uses resources wisely to preserve them for the future

Thinking people sense that wiser approaches to resource management are required, and they buy into the concept of sustainable development. While this may sound like an academic concept to some people, it is really a practical imperative. It may, however, increase political challenges to water managers, because customers expect *both* reliable water supplies and environmental protection. In fact, water industry research already shows that it is good policy to emphasize environmental stewardship as a business strategy.

To practice TWM, tomorrow's water industry will require utility and government leaders as well as regulators who can span the needs of the economy and the environment. Policy makers will need keen insight into the water industry and its incentives, and citizens will have to practice greater stewardship. The knowledge industry, including consultants and researchers, will have to provide new ideas, and vendors must create new products. Above all, leaders who can be "master integrators" will be required to become effective public managers in the twenty-first century.[1]

This book serves as an instruction manual for these master integrators. It is about the balance between our responsibilities to provide safe and reliable water services and to protect the environment. The chapters address the management of water resources rather than specific issues of water treatment or distribution.

When the American Water Works Association (AWWA) developed the TWM concept, the focus was on water supply services. However, TWM's definition shows that it applies to all water services—supply, wastewater and water quality, agricultural water, hydropower, instream flow management, and security against flood losses. In other words, TWM goes beyond narrow definitions of water management to total water stewardship. It is a term, similar to Integrated Water Resources Management (IWRM), that describes taking an overall approach to solving water problems.

---

[1] Donald Stone (1974) saw the need for integrative problem solving through the field of public works management and Joseph Bordogna (1998) saw it through education and research.

TWM is about leadership

At the end of the day, TWM is about leadership. Given this, the question of "Whose point of view?" becomes critical. Are we focused on a utility serving its customers or on the needs of the broader society? The answer is, we focus on both. This is clear from the definition of TWM: "stewardship of water resources for the greatest good of society and the environment."

Can TWM serve both the environment and society? Is what's good for General Motors also good for America?[2] It will have to be. TWM requires participation of utilities, business, and government. As Figure 1-2 shows, business and utilities are pulled in different directions but in different ways. One way is to make a profit or be a successful enterprise. The other is to reach out to handle social responsibilities.

TWM is clearly in society's best interests, but what are the incentives for utilities to embrace it? This fundamental issue creates a clash of culture that is captured by the phrase "it's not my problem." TWM requires that incentives be created. Otherwise, TWM will be just a visionary concept with little practical value. The key is to move past vision and on to action.

Water managers know that, above all, they must provide reliable, safe, and secure water services to their customers. This imperative trumps all others because it is their direct responsibility. If confronted by a value set that threatens this responsibility, the direct mission will come first.

> **TWM must be more than visionary; its challenge is to move from vision to action**

It is hard to share power and to say to someone with a different value set, "Let's cooperate to solve our problems together." The result is a system that is more adversarial than cooperative. This leads to the "it's not my problem" syndrome, which says "Don't bother me with that," or, at worst, "We'll see you in court." One of the institutional problems that confront TWM is that some people benefit by keeping the adversarial process going. These types of problems are discussed in chapter 10.

The real challenge is to meet direct needs of your organization *and* to work with others to meet their needs, too. The military analogy explains why this is difficult: it's much better to be on one side of the valley or the other, because the people trying to balance things from the valley floor get shot at from both sides!

---

2 This quote, attributed to General Motors president Charles Wilson in 1955, suggested that society's best interests were the same GM's. For TWM, the comparable question is, Is what is good for utilities good for society?

6   TOTAL WATER MANAGEMENT

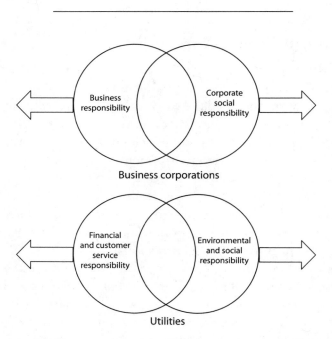

Figure 1-2.   Financial and outreach responsibilities of business and utilities

To some water managers, sustainable development sounds like an environmental ploy to get more concessions from an industry that already has difficult problems in delivering reliable and safe water services to its customers. Most challenges are in the political rather than technical arena. While politics vary, sustainable development and the need to manage water wisely are shared values around the world, transcending geography, culture, and religion. People share values such as good public health, environmental protection, and a fair deal for everyone. Every person, animal, and plant on the globe depends on water for life. So the central dilemma is how, with growing populations and demands for a limited resource, do we manage water so all needs are met without spoiling the environment for tomorrow?

There are plenty of slogans for meeting today's needs without spoiling the environment, but the challenge is to make them work. Without its defining principles, TWM could be just another one of those slogans. Its definitions and principles (see chapter 3) tell us what it is, but how to practice total water management requires the explanations given in the remaining chapters.

As an idea, TWM captures our imagination about addressing issues and stakeholder needs. John Young (2006), chief operating officer of American Water, wrote that it is to "assure that water resources are managed for the greatest good of the people and environment and that all segments of society have a voice in the process."

Taken together, TWM and related concepts such as Integrated Water Resources Management form a dominant paradigm that is legitimized by professional organizations, the media, government agencies, educators, or other mechanisms (Wikipedia, 2006).

Positive practices meet the needs of the present without sacrificing resources for future generations. Figure 1-3 shows the balance point for sustainable development. However, these positive practices are difficult to implement. Unsustainable practices are negative and harmful to nature.

Is there an environmental crisis?

Is there really an environmental crisis that requires TWM, or is this invented by radical groups and the media? To answer that, think of three groups of people. One thinks that pollution, drought, and waste of water are bringing global disaster. A second group is busy solving the practical problems of supplying and managing water. To the third group, access to clean, safe, and low-cost water on demand is more important than the debate about the environment. For convenience, let's call the three groups the "environmentalists," the "water managers," and the "citizens."[3]

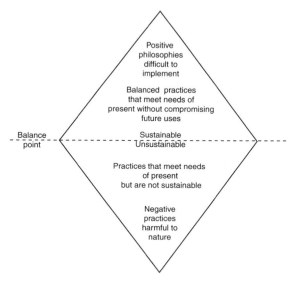

Figure 1-3. Balance point for sustainable development

---

3 These terms do not minimize the views of any group mentioned here. Use of the terms will facilitate discussion about environmental goals, practical water management issues, and citizen rights and responsibilities.

The environmentalists keep major issues in front of us, such as global warming, loss of forest and wetlands, and extinction of species. Lest we say that they are too global, they also skirmish in local places about small losses such as paving over wetlands. They perform a service to society in keeping our attention on environmental issues.

Water managers are concerned about environmental issues, but they focus on the immediate problems of managing infrastructure, raising revenue for operations, recruiting skilled workers, and complying with regulations.

As citizens, we are all environmentalists to some extent. However, we have different views of issues and what to do about them. Probably most agree that unless water is managed better, both the environment and society will suffer. We part ways in deciding how to manage water better and in assessing how the suffering will occur. Most of us would admit that many people are not tuned in to the water conversation. People focus on things that interest or concern them and do not get involved in every issue.

The positions of the three groups can be summarized by saying that the environmentalists push a sustainability agenda, water managers want sustainability but are focused on their direct missions, and most citizens do not tune in to the conversation very much. This triangle of groups creates a TWM balancing act, as shown in Figure 1-4.

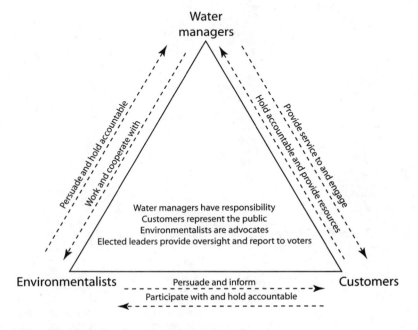

Figure 1-4.   The balance in water management

Barriers to sustainability

Although TWM provides a powerful platform to place water management on a sustainable use basis, it faces formidable challenges. The challenges arise from myriad small impacts that cause problems, not a single large foe to conquer. The water crisis is also a creeping crisis, easy to ignore in the short term. It is not a massive, sudden crisis that galvanizes a lot of political support.

Water managers must be leaders in TWM, but they do not work with only a few partners. They must work broadly with stakeholders and citizens in many situations they do not control. Utilities and regulators work in an ordered business world, but many small and seemingly unorganized players also affect water systems.

The shared problem of water managers is to provide sustainable water management services for human and economic needs while maintaining the environmental quality that underlies economic prosperity. Sustainability requires balanced water supplies for humans and the environment, protection of water sources, and resolution of water conflicts at scales from local to global. Solutions must include resource sharing, governance systems, and reduction of hazards. Water scarcity requires new technologies for water efficiency, conflict is mitigated by knowledge of shared benefits and improvements to the Triple Bottom Line (TBL),[4] and vulnerability requires improved security against natural and human-caused threats.

To implement these solutions requires substantial societal efforts in the face of formidable challenges. Moreover, they will defeat us unless we sustain actions on multiple fronts against many small challenges, while being vigilant and not allowing a creeping crisis to overwhelm us.

Who is in charge of this shared problem? No one really manages the myriad of smaller actions that fly under the radar screen of the water industry, and no one is in charge of finding solutions to the shared problems. That is why, from time to time, someone will say, "We need a czar to take control of this water issue." The European Union's Water Framework Directive calls for a leader on water issues in the form of a "competent authority" (Green and Fernández-Bilbao, 2006). This is, to some extent, another call for a czar, but there won't be one in the United States because the public doesn't generally support more government oversight and instead seeks decentralized and private-sector solutions. Water utilities on the one hand and business and private citizens have to work together

---

[4] The metaphor of the Triple Bottom Line originated in the sustainability movement and refers to accounting for economic, social, and environmental costs and benefits. It will be explained in chapter 5.

to solve the problems. We have to create through our shared work an "invisible hand" to solve the problems, albeit with some badly needed coordination.[5]

The Tragedy of the Commons: people care for their own property but not the property of others

The water industry manages big systems, such as dams, large diversions, and discharges from wastewater treatment plants, but innumerable smaller actions caused by land management activities are not under the direct control of water utilities. They mainly involve nonpoint sources and hydrologic modifications, two activities originating in the broader society. Also, many small storage, diversion, and discharge actions are also caused by small players who do not fall within the spheres of influence of the large utilities and agencies. So we think of TWM as mainly an activity of water managers, but one that also requires broad engagement of society to deal with the myriad of small impacts that affect water systems (Figure 1-5).

The nature of TWM

TWM is a systemic concept, much as is shown in Figure 1-5. This illustrates how industry and citizens have roles in directing water uses toward positive contributions to the economy, the environment, and society. The Triple Bottom Line is the way to keep score (see chapter 5).

Principles and practices

Given the broad scope of TWM, it is a challenge to create a clear set of principles and practices to define it.[6] However, without a defined set of principles and practices, the concept of TWM remains ambiguous. It is based on notions that may seem soft, such as stewardship, shared governance, coordination, and conflict resolution; nevertheless, TWM can be implemented through specific actions and processes.

So, it is natural to ask again, is TWM for real or is it only fantasies and dreams? Is it simply another visionary concept that is doomed to an early death as soon as another trendy phrase comes along, or does it have practical value? In many ways, the jury is out on these questions because of

---

5 The invisible hand metaphor was made famous by eighteenth-century economist Adam Smith, who wrote *The Wealth of Nations*. In water management we need an invisible hand that occurs from the shared work and coordination of water industry players who make the system work in spite of the lack of centralized government control or marketplace solutions.

6 Chapter 3 presents a definition and set of principles and practices for TWM.

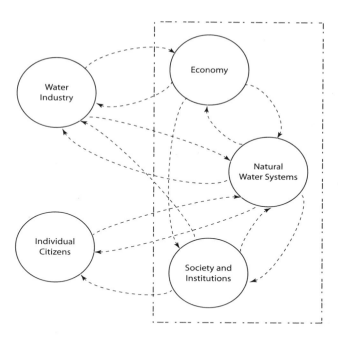

Figure 1-5.   Total water management: a systemic concept

the institutional barriers to it (see chapter 10). But like in any challenging area, moving forward is progress.

The institutional barriers are driven by the "it's not my problem" syndrome, in which players take decisions and actions in their own interests and create a version of the Tragedy of the Commons, the phenomenon of people caring for their own property but not the common property of the public (see chapter 10).

Obviously, water management would work better if everyone thought about others' needs as well as their own, that is, there existed a Golden Rule of water management. However, this fantasy quickly founders on the shoals of reality, and incentives and penalties are needed to encourage or force people to do the right thing in water actions. Everyone may agree that TWM is a good idea, but when it comes to spending money or giving up political power for it, they will balk and say, "That is not my problem."

Laws and regulations

Sustainable water management will occur only if all stakeholders take decisions and actions that align with sound principles, requirements, and incentives. However, utopian concepts and changed attitudes alone will not make people do the right thing. For ethical and sustainable water management, regulatory controls and law enforcement are required.

Laws and regulations compel stakeholders to meet minimum environmental and health rules, but if the players do only the minimums, water systems tend to degrade. The consequence of this drive toward minimal standards is that developing regions will experience unsustainable development that causes water scarcity, conflict, and vulnerability. So regulations are necessary but also not sufficient for TWM to occur. This is particularly the case with all the activities and policies that affect water resources.

Stewardship goes beyond regulations

The "it's not my problem" syndrome is a natural consequence of the fact that water utilities and organizations have challenging corporate responsibilities that require their full attention, even before they worry about societal issues and stewardship of the environment. They give their primary attention to meeting the budget, building facilities, delivering services, and avoiding regulatory sanctions.

In addition to this situation of benign neglect, society also faces challenges caused by greed, incompetence, malfeasance, and ignorance. If you take this attitude to the limit, you end up with a situation in which everyone withdraws inside his or her own corporate castle and a lot of the territory is left unguarded, subject to raids or just general neglect.

Good policy is needed to get organizations and individuals to care for the public space as well as their own spaces. While they agree about the need for the broader societal responsibilities implied by TWM, the challenge remains how to get it done.

TWM—more political than technical

While the definition of TWM emphasizes principles and practices of water management, at its center are ideas about economics and politics. The core economic issue is a search for the greatest good of society and the environment, which is a general goal of public-sector economics anyway. The TWM definition includes the subsidiary economic goals of efficient allocation of limited water resources to address social values, cost-effectiveness, and environmental benefits and costs.

The political ideas of TWM are that stewardship is a public responsibility and that TWM requires participation from all units of government and stakeholder groups to balance competing uses of water in spite of local and regional variations and issues. If successful, this coalition will foster community goodwill and public health and safety. The political statements outline how the process should work and what it should achieve.

You need look no further than the general welfare provisions of the US Constitution for authority for these economic and political ideas. The Preamble states: "We the People of the United States, in Order to form a more perfect Union, establish Justice, insure domestic Tranquility, provide for the common defense, promote the general Welfare, and secure the Blessings of Liberty to ourselves and our Posterity, do ordain and establish this Constitution for the United States of America." Article I, Section 8 states: "The Congress shall have Power to lay and collect Taxes, Duties, Imposts and Excises, to pay the Debts and provide for the common Defense and general Welfare of the United States" (US Government 2006).

TWM's focus on the greatest good of society and the environment is aimed directly at improving the general welfare, and stewardship and balancing competing uses are necessary conditions to achieve the greatest good.

Two powerful philosophical ideas also support TWM: environmental ethics and corporate social responsibility (see chapter 11). Environmental ethics is the study of our right behavior toward the environment. The argument is that, as a public good, water has been bestowed on people and nature (the environment), and it is our responsibility to care for it.

Stewardship is closely related to the goal of corporate social responsibility (CSR), meaning the responsibilities businesses have to make contributions beyond the profit motive (Hay, Stavens, and Victor, 2005). How do firms balance their fiduciary responsibility to shareholders with CSR? It can be argued that any firm, agency, or organization that affects natural water systems has a public responsibility to care for water because of water's shared uses, including environmental uses.

So TWM is a paradigm for water management to work effectively in a democratic political system, with the rule of law and a mixed public and private economic framework that underlies the political framework. As the "art of government, politics is important for governance to occur and for negotiating conflicts and balancing outcomes to meet goals and objectives within the economic framework.

Water has a high political intensity because people have different agendas that are worked out in the political process. Politics and governance provide a set of rules and processes to resolve differences and make positive things happen.

The difference between what people are required to do and what they ought to do is the difference between law and stewardship, or social responsibility. What we ought to do is governed by rules that fit within social norms and are part of the institutional fabric of the water industry.

Use of case studies to explain TWM

Sometimes a case study helps us see the interplay of issues in water management situations. A case study is like a story with a setting, characters, action, and a conclusion. Case studies add experience-based learning to expository information. They raise interest levels, show how decision systems can make the difference between success and failure, and give a sense of participation in real-world political situations. This book uses brief case studies and examples of interactions and water decisions to illustrate the principles and practices of TWM.

Water management is a shared challenge

The case-study method involves analysis of complex situations requiring remedial treatments of some kind, as in medicine, law, or business situations. The first case at the Harvard Business School was in 1912, and by 1924 the case method had been adopted as the primary method of instruction (Ewing, 1990). Also at Harvard, the John F. Kennedy School of Government (1992) uses the case method to explain decision-making in public administration.

Action-forcing cases place the reader in the shoes of government officials faced with a problem requiring action and ask, "What would you do?" Retrospective cases tell the whole story, including the decision and the consequences (Kennedy and Scott, 1985). A good case is short but general, has pedagogic utility, and is conflict-provoking and decision-forcing (Robyn, 1986).

Just as in complex law or business cases, water problems are amenable to explanation with case studies. Chapter 2 offers a detailed case study in the form of a story to illustrate how players in water management interact with each other. Other examples throughout the book offer fragments of case studies.

What does the book contribute?

This book explains TWM's goals and principles and outlines the institutional challenges to making it work. It offers policy prescriptions for overcoming the main challenges. It explains how the water industry works and how decisions to control water resources systems are made. The book also explains how myriad small actions fly under the radar screen of the water industry but are important to sustainable water management, and defines TWM and its elements and how they work within the water industry.

The next chapters examine the important concept of shared governance and how water actions are evaluated under the Triple Bottom Line concept of sustainability. Chapters 9 and 10 explain the political and legal forces that shape how the water industry works. In this highly regulated industry, water service providers operate under the close scrutiny of regulators and with participation by a public and private support sector. The book also discusses how the myriad of small actions work in a different legal and political environment.

Given that "beneficial human and environmental purposes" involve economic, environmental, and social impacts, chapters 6, 7, and 8 are devoted respectively to explaining how these are evaluated.

The final chapters provide an analysis of how to make TWM work better in the highly regulated and political water industry and how environmental education can improve TWM. The concept of institutional arrangements is used as an organizing concept to describe how incentives, roles and relationships, and controls on the industry can be used to make it work better.

In summary, the book explains the following:

- How TWM is a formidable challenge, and while utilities cannot take all the responsibility, they can lead in promoting shared governance and corporate social responsibility toward sustainable water management;
- The problems that create a creeping water crisis, including mega-issues such as dams, diversions, and discharges and the many hydrologic modification and nonpoint source issues;
- The principles and practices of Total Water Management and how creating a workable institutional framework for them is essential to achieve balanced environmental management;
- Roles and the circle of responsibilities and how to deal with the "it's not my problem" syndrome;
- Requirements of sustainable water management, including big systems as well as impacts of small water systems and hydrologic modifications;
- How water is a creeping crisis, not a sudden one, and that public awareness and support for the full value of water to society are needed, as well as the critical roles of environmental education and sustainability training;
- Tools for Triple Bottom Line reporting, along with economic, environmental, and social impact assessment;
- Principles of shared governance, regional cooperation, and river basin planning as they apply to TWM; and
- How to promote social harmony and community spirit by linking environmental ethics, citizen responsibility, and stewardship to water management.

Figure 1-6 shows the elements of TWM and how they are addressed in the book.

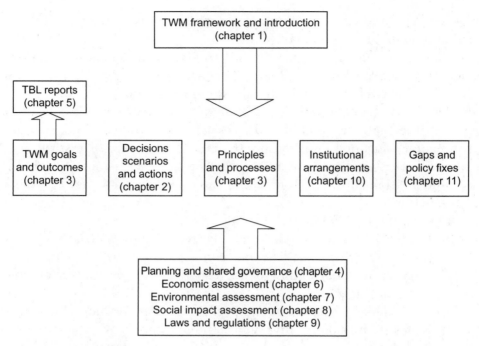

Figure 1-6.　How TWM works

How do utilities take the lead?

Given the difficulty of managing an enterprise in the first place, how could a utility do other than take care of its own customers? How could it focus on the broader public good and still perform its mission? It is not a perfect world, and no one can shoulder society's whole load.

Actually, this question is at the heart of the long debate over the role of government in the economy. In a communist system, benefits and costs are to be spread over society, and in a dictatorship, a wise authority is to decide how to do that. In reality, neither works and we revert to democratic capitalism, which has competition and an imperfect political system. So we have to make it work as best we can.

That's why utilities need to take a lead role in TWM. Utilities are the logical leaders for much of the work involved with TWM. They have more resources than other institutions, and they have experts who know about and care about water. However, utilities are not alone in this fight because there are plenty of leadership roles to go around.

Summary points

- TWM is a framework for principles and practices of sustainable water resources management. It is a way to work on water problems within a democratic political system with the rule of law and a mixed public and private economic system.
- Evidence shows that TWM is needed because the capacity of the environment to bear the load may be nearing its limit. Sustainable development and management are TWM concepts for using resources wisely to preserve them for the future.
- Large water issues are controlled by permits and other government actions, but smaller actions and shared problems are more difficult to manage.
- Water utilities and organizations have challenging responsibilities that require their full attention, and it is sometimes difficult for them to focus on shared societal issues such as sustainability. Moreover, environmental issues are perceived differently by environmentalists, water managers, and citizens. Utilities should take a lead role in TWM because they have more resources than other institutions and they have experts who know about and care about water.

Review questions

1. Define sustainable water resources management and explain how TWM fits within it.

2. Why have visionary concepts such as TWM not been embraced more widely by the water management community?

3. Which stakeholder group(s) should have the major responsibility for TWM? Why?

4. If TWM adds cost to a water utility's operations, who should bear that cost and how should it be financed?

5. Give examples of small actions that degrade water resources but are not controlled readily by government regulators.

References

American Water Works Association. 1994. White Paper: Principles of Total Water Management Outlined. *Mainstream* 38 no. 11: 4,6.

American Water Works Association Research Foundation. 1996. *Total Water Management Workshop Summary.* Draft. Seattle, Wash., August 18–20. Denver, Colo.: AwwaRF.

Bordogna, Joseph. 1998. Tomorrow's Civil Systems Engineer: The Master Integrator. *Jour. Professional Issues in Engineering Education and Practice* 124 no. 2 (April): 48–50.

Ewing, David W. 1990. *Inside the Harvard Business School*. New York: Times Books.

Green, Colin, and Amalia Fernández-Bilbao. 2006. Implementing the Water Framework Directive: How to Define a "Competent Authority." Universities Council on Water Resources, *Jour. Contemporary Water Research & Education* 135 (June): 65–73.

Hay, Bruce L., Robert N. Stavins, and Richard H.K. Vietor. 2005. *Environmental Protection and the Social Responsibility of Firms*. Washington, D.C.: RFF Press.

John F. Kennedy School of Government. 1992. *The Kennedy School Case Catalog*, 3rd ed. Cambridge, Mass.: Harvard University Press: p. i.

Kennedy, David M., and Ester Scott. 1985. *Preparing Cases in Public Policy*. Boston: Kennedy School of Government, Harvard University (N15-85-652.0).

Postel, Sandra. 2007. Aquatic Ecosystem Protection and Drinking Water Utilities. *Jour. AWWA* 99 no. 2: 52–63.

Robyn, Dorothy. 1986. "What Makes a Good Case?" John F. Kennedy School of Government. Cambridge, Mass.: Harvard University Press.

Stone, Donald C. 1974. *Professional Education in Public Works/Environmental Engineering and Administration*. Chicago: American Public Works Association.

US Government. 2006. Constitution of the United States. http://www.house.gov/paul/constitutiontext.htm. Accessed October 19, 2006.

Wikipedia. 2006. *Paradigm*. http://en.wikipedia.org/wiki/Paradigm. Accessed September 28, 2006.

Young, John. 2006. Challenges and Benefits of Total Water Management. *Jour. AWWA* 98 no. 6: 32–34.

CHAPTER 2

WATER MANAGEMENT AND ITS IMPACTS

Water managers face many types of issues, and to be specific let's place TWM issues into three categories:
- The *water supply problem*: Maintaining access to water sources without damaging the environment,
- The *water quality problem*: Sustaining and improving water quality for customers and the environment, and
- The *environmental problem*: Avoiding degradation from nonpoint sources and hydrologic modification.

The many water management problems that fall into these categories are like the pot slowly brought to a boil that kills the unsuspecting frog immersed in it. They are shared problems that we can't always control. Utilities have big roles in solving them, but they seldom can act alone. The critical issue is to get everyone to pull in the same direction.

Why should players in the water industry pull together? Are the issues serious enough to bypass "business as usual" and come up with creative approaches? The consensus is that they are, and this chapter explains the issues and the water industry's roles in confronting them. It identifies the players, officials, and support groups, as well as the dynamics between water use and the environment. It explains the large and small actions that affect supply, water quality, and the environment. To illustrate the many issues involved, the chapter concludes with a case study in the form of play.

The water supply problem

Water supply is the core service of the water industry, which is changing from a centralized, slow-reacting, supply-side industry to a distributed, flexible, and faster-acting one that practices demand management. This creates new demands on the water manager, including the ability to react more quickly, respond to more constituents, and manage more complex systems.

The main water supply problem that utilities face is sustaining and expanding an adequate and high-quality source to meet customer demands. For some utilities with abundant sources and low population growth, this has not been a significant problem, but many others face supply and growth pressures at the same time.

In a supply-driven paradigm, water utilities provide water under a cost-of-service pricing approach. This approach does not conserve water, and the shift is toward demand management with more emphasis on conservation and pricing. Pricing can foster water-use efficiency, both in urban and agricultural settings. Pricing integrates many issues, and those who take the private-market view, as opposed to the public-good view, tend to favor an approach to setting water rates where they are used to control water use, as opposed to a cost-of-service approach, which only reflects the cost of the water and not the demand for it.

Available potable water supplies are getting harder to find, but the good news is that although just a few years ago water use in the United States was increasing, total use has stabilized since about 1985. Groundwater withdrawals during 2000 were 14 percent more than during 1985, but surface water withdrawals were stable.[1] These macro figures hide the fact that gaining access to expanded and more reliable supplies is becoming much harder for utilities than it was in the past.

Water pricing and demand management go together

High-income modern societies use more water than lower-income traditional societies for such things as lawn irrigation, air conditioning, microchip production, and power generation. The increased uses have created scarcities where none existed before. As scarcity increases, the resource should be allocated to the most productive and valuable needs, and this is where debates over the "most valuable" uses of the water occur.[2]

---

[1] Water use is reported in USGS (Hutson, Barber, Kenny, Linsey, Lumia, and Maupin, 2004) and by water industry surveys, such as AWWA's Water://Stats database.

[2] See chapter 6.

Clearly, water-demand patterns have changed. During the past 50 years, irrigation use peaked, mining use declined, industrial uses became much more efficient, and household water conservation became common. In the future, water will cost more than in the past, and more conservation will occur. Also, there will be a continuing reallocation of water from low-value uses to higher-value uses.

The water quality problem

The water quality problem is even more vexing than the water supply problem. Because the concept of water quality is so general, no agreement exists about a single index for it. Not only is the concept very general; there is much disagreement about how it affects ecosystems and human health, so the regulatory processes that set standards must be carried out with care to make sure all stakeholder views are included.

In some ways, water quality in the United States is stable, but in other ways it is worrisome. Point source discharges from a single pipe or channel seem to be mostly controlled, but nonpoint sources where the contaminants are dispersed on the land and carried to streams by rainwater are worsening. Examples include any runoff from urban areas; rural lands, including cropland and forests; seepage from adjacent groundwater sources, highways, and airports; and other land uses.

We cause our own water quality problems. A study called Water Quality 2000 (1992) explained that pollution is caused by how society lives, farms, produces, and consumes; transports people and goods; plans for the future; and acted in the past. The study, also called WQ 2000, listed nine sources of water contamination, which are described in Table 2-1.

Issues that have emerged since WQ 2000 include pharmaceuticals in water, algal toxins, and distribution system water quality. A 2003 report by the Natural Resources Defense Council (NRDC) rated drinking water in some US cities as poor, with trace contaminants showing up in tap water. The ability to measure more constituents, including pharmaceutically active components, has improved, heightening our awareness of foreign substances in the water even further (Singer, Doherty, and McMullen, 2003).

Public concern about water quality is growing. Aging customers are more vulnerable, especially patients undergoing chemotherapy and the transplant population. The public wants resolution of these issues and is taking health into their own hands. Sales of bottled water and home treatment devices have increased and have significant growth potential.

Table 2-1. WQ 2000 sources of water contamination

Source	Effects
Agriculture	Farm operations discharge large volumes of sediment and nutrients and smaller quantities of toxic chemicals. Accounts for wetland losses and damage to riparian and floodplain environments. Runoff from animal production is a source of phosphorus and pathogens in lakes, and agricultural chemicals threaten groundwater.
Atmospheric sources	Acidic or toxic substances may be deposited in lakes or estuaries. This may impair aquatic ecosystems, cause algal blooms, and even be lethal to aquatic organisms.
Community wastewater systems	Treatment plants remove much contamination, but they face challenges to maintain effectiveness. They may not remove toxic substances, may miss nonpoint sources, and may be bypassed by combined sewer overflows.
Industrial dischargers	While industries are generally in compliance with permits, they discharge a massive quantity of conventional and toxic substances along with thermal pollution.
Land alteration	Logging, mining, grazing, and land development change runoff and add sediment and chemicals to the water. They may also destroy wetlands and habitat.
Stocking/ harvesting of aquatic species	These may impact aquatic ecosystems.
Transportation systems	Ships, roads, rail, and pipelines impact the waters. Oil spills are a major source of contamination. Transportation may destroy habitat, as, for example, through dredging.
Urban runoff	Similar to land development, this causes contamination through release of sediment, organics, oil, and toxic chemicals.
Water projects	Water projects may reduce habitat through channelization, dams, and consumptive use of water, and may impact anadromous and riverine fish.

Source: WQ 2000, 1992.

The environmental problem

The environmental problem is caused by decisions about water that hurt the environment and its living things. The concept of the *water environment* helps us identify the zones around streams and lakes where aquatic life and health are important to fish and wildlife and to scenic values. We can think of water in these zones as nourishing soil systems, plants and biota, microbes, insects, fish, and wildlife.

These "customers" for water have requirements that are measured by instream flow rate, volume, timing, temperature, and by the physical, chemical, and biological qualities of water. Water quantity changes, whether in rate, volume, or timing, lead to hydrologic alteration. Changes in water quality impact habitat and living organisms directly.

The impacts on the environment from water management changes can be either positive or negative. A few examples include:
- Fish habitat changes
- Introduction of exotic fish species
- Bird habitat changes
- Changed stream widths
- Toxics in sediments
- Stream food-chain interruption
- Wetlands destruction
- Estuary deterioration
- Mussels
- Groundwater changes (subsidence, lowered water tables in wetlands)

The environment is a water customer, too

The Tragedy of the Commons

TWM addresses water supply, water quality, and environmental problems at the same time. However, many issues fall between the cracks because of faulty institutional arrangements, which are the forces that cause people to act the way they do—laws, customs, incentives, etc. This is a finding of WQ 2000: How we live and act determines water quality, and it also determines our ability to create sustainable solutions.

Why problems fall between the cracks is explained largely by the metaphor of the Tragedy of the Commons, which has endured as a useful slogan. The term goes way back in time, but it was popularized by Garrett Hardin (1968) and has stuck since then (Wikipedia, 2007). It means that people take actions in their own interests and care for their own property but not for the common property of the public. The commons refers to common property, like a city park, but back when the term was first used, people let their animals graze on the commons.

The Tragedy of the Commons applies to organizations as well as to individual actions. In organizations, people are rewarded for advancing the interests of their company or agency, thus they may take the "it's not my problem" stance so they can focus on the bottom line or their agency's mission. That's where corporate social responsibility comes in as a second mission of business, as a complement to profit.

Figure 2-1 illustrates how, when agencies look inward and concentrate only on their core missions, a lot of territory goes unattended, but if they also reach out with their actions, the territory is better covered.

Without the right incentives, neither individuals nor organizations will embrace TWM. Why institutional arrangements can be faulty is explained in chapter 10, and ideas about how to overcome the gaps through

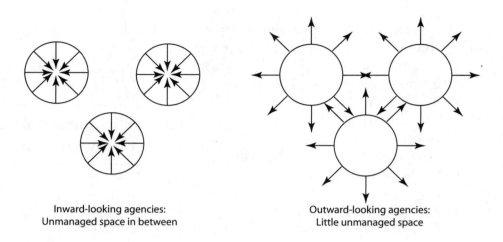

Inward-looking agencies:
Unmanaged space in between

Outward-looking agencies:
Little unmanaged space

Figure 2-1.   TWM looking inward and outward

improved stewardship are presented in chapter 11. This chapter focuses on TWM's arena of action and the roles in which it is required. It explains how stewardship of water resources is not only the responsibility of water utilities but of all citizens.

Arenas for action of TWM

If the limited supply of water is to be managed on a sustainable use basis, then all uses that affect sustainability must be considered. This arises from the TWM definition: "A basic principle of Total Water Management is that the supply is renewable, but limited, and should be managed on a sustainable use basis." This means that the arena for action for TWM embraces all water uses.

Also according to its definition, TWM will respond to the sustainability challenge if it

- balances competing uses through efficient allocation by planning and managing dynamically;
- adapts to changing conditions and local and regional variations; and
- uses coordination and conflict resolution to reach decisions, with participation of all units of government and stakeholders.

These practices are essential. To implement them, shared governance is required, because managing the shared uses of the "commons" is required. This helps in larger water management actions, but the myriad smaller actions are more difficult to handle.

CHAPTER 2   WATER MANAGEMENT AND ITS IMPACTS   25

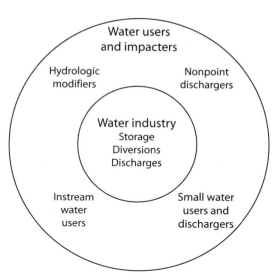

Figure 2-2.   Water industry and impact sources

Water management actions involve different types of uses, or any reliance on it for some purpose, whether for supply, to convey wastewater, or to avoid flood damages. Thus, any action that affects water quantity (meaning rate or volume) or water quality becomes a use of water and is of interest to us.[3] Figure 2-2 shows how these impacts occur from the water industry itself and from many other sources, large and small.

It is important to recognize that water utilities do not control all of these water uses. In addition to water diversions and discharges, various land uses affect water quality and quantity. Academics lump these by referring to how water and related land uses affect natural water systems.[4]

While there are many examples of water management actions, you can put them into five groups. Table 2-2 provides a list of direct uses of water and land uses that affect water.

By acknowledging that TWM covers management of water through storage, diversions, and discharges, as well as nonpoint sources and hydrologic modifications, the need for shared responsibility is made clear. The focus in TWM on watershed management is appropriate, because water managers must protect against overstressing streams, prevent nonpoint sources, and take other necessary actions to safeguard water supplies.

---

3   Nonpoint source control illustrates this point. Say a land developer builds a parking lot and it pollutes a nearby stream with runoff. Is this a use of water or is it an indirect impact on water from the land development activity?

4   Water and related land uses are combined in many policy studies and books. See for example Black and Fisher, 2001.

Table 2-2. Management actions that impact water resources

Actions	Examples
Water storage. Large including dams impounding more than 50 acre-ft and smaller including ponds and small lakes.	Some 78,000 dams and any significant storage for various purposes.*
Diversions. Any diversions or pumping of water from a stream or aquifer, both temporary small-scale and large permanent constructed diversions.	Diversions for any water supply. Includes small, identifiable diversions for industries, businesses, or farming.
Point source discharges. Any discharge from a wastewater treatment plant (WWTP) or point, such as stormwater drainage, where significant quantities of water are discharged.	Discharges from some 17,000 WWTPs and identifiable industrial or commercial dischargers, and discharges from large package plants and individual treatment systems.
Nonpoint source discharges. Any discharges to streams other than point discharges, including very small and diffuse discharges and covering both large and smaller areas.	Small water discharges that cannot be readily identified, including smaller individual treatment systems, stormwater systems, farm return flows, and industrial wastewater systems
Hydrologic modifications or alterations. Any stream disturbance, constriction, fill, or alteration of stream. Refers to altered quantity of runoff, whereas nonpoint discharges refer to quality of altered water.	Small ponds, culverts, filling in of stream, channel changes, stream maintenance, wetland disturbance, road construction, and drainage.

* This count is from the inventory by the US Army Corps of Engineers included in the National Inventory of Dams, see ASDSO (2004).

To get a better idea of the differences between large actions, where utilities logically have responsibility, and smaller actions that lie outside their purview, Table 2-2 includes a range of actions, between large dams and small ponds, between large and small water systems, and between organized treatment systems and individual waste treatment units. The number of small actions is many, many times that of large actions and they are difficult for anyone to control.

These large and small impacts on water resources are summarized in Figure 2-3.

Why sustainability is a shared responsibility

Does TWM include only direct functions such as diverting water and returning it to the stream, or does it also include indirect functions such as land use, stream modification, and nonpoint source control? It is easy to see that sustainable development must include both and be a shared responsibility across society, but how do people participate in it via TWM? The answer lies in how water management actions work to affect water

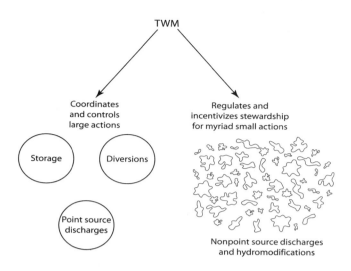

Figure 2-3. How TWM relates to large and small actions

as it passes through the hydrologic cycle, is used, and is discharged to the environment again.

A 1996 Awwa Research Foundation (AwwaRF) workshop group[5] answered the question by defining TWM as the "exercise of *stewardship* of water resources for the greatest good of society and the environment" (emphasis added). Stewardship means caring for something because you are responsible for it. Caring for water resources requires control of land uses, stream modifications, and nonpoint discharges, as well as diversions and point source discharges. Therefore, TWM must cover all water management functions and be a shared activity of society, not limited only to the direct roles of water utilities.

How society organizes this shared responsibility for stewardship of water is a central issue of TWM. Utilities have specific and direct responsibility for water management, but citizens have responsibility, too. At one level, responsibility is expressed through the regulatory structure that controls how water is managed, but these regulations alone are not enough. They work fairly well for large water users or dischargers, but many small actions elude regulators. Even when they control permit conditions, meeting the minimums is often not enough.

---

[5] I participated in the workshop and was impressed by the knowledge of the some 30 professionals who prepared a concise but meaningful definition. Their starting point was the White Paper published by AWWA (1994; see Appendix A).

In practical terms, the many small actions are hydrologic modifications and nonpoint sources, as well as any other diversion, storage, or wastewater discharge that occurs. Stewardship is everyone's responsibility, but that can mean it is also nobody's responsibility. The theory is great; getting the job done is where the rubber meets the road. Roles and responsibilities to get the job done are discussed in detail in chapter 10.

Threats to sustainability

The actions that threaten a sustainable environment can be illustrated by a risk triangle, which has threats at one corner, with the other two corners showing vulnerabilities and consequences, respectively.

Threats to natural water systems come from storage, diversions, discharges, nonpoint source discharges, and hydrologic modifications. Table 2-3 details these threats along with ways to mitigate them. As Table 2-3 suggests, if the mitigation measures are taken, the vulnerabilities and the consequences will be less, even if the threats remain.

Water utilities are aware of a wider range of risks, of course. As Figure 2-4 shows, they face dual risks. One is the risk to natural water systems for which they are accountable, and the other is risk to their own businesses.

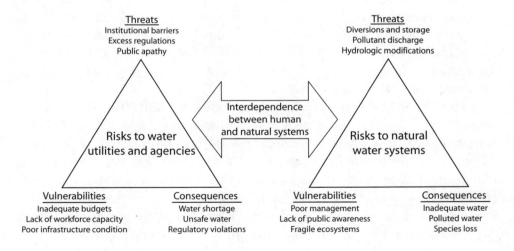

Figure 2-4. Dual risks of water utilities

Table 2-3. Threats to natural water systems

Threats	Mitigation strategies
Larger and more identifiable actions	
Water storage: changes the ecology of streams, captures sediment, leads to downstream erosion, alters water quality, and flattens natural hydrographs.	Mitigation for storage maintains release hydrographs as close to natural as possible and avoids construction of reservoirs when possible.
Diversions: dry up streams, alter natural flows, reduce stream cleansing power, and degrade ecosystems.	Diversions should be minimized. Instream flow programs and permits to limit withdrawals are required.
Point source discharges: from treatment plants change water quality.	Effective enforcement of the Clean Water Act helps to mitigate pollution impacts.
Smaller and dispersed actions	
Nonpoint source discharges: create large and cumulative impacts on stream water quality and sedimentation.	Best management practices are the way to mitigate nonpoint sources.
Hydrologic modifications or alterations: stream disturbances destroy ecosystems, add sediment, and degrade natural channels.	Hydrologic modifications should be minimized and best management practices should be used.

Players and the water management actions they control

The players who control the actions that affect water resources start with the formal members of the water industry: water service providers and regulators. They also include the public at large and other organizations that impact water systems. If all of these players are not engaged, TWM will fail. In other words, no one can go it alone.

The people and organizations that affect water resources but fall outside the world of organized utilities, industries, agencies, and regulators are hard to categorize. Table 2-4 provides a list of them along with a short description of how they impact water systems. Some overlap will occur in the categories.

If we add the formal water service providers and place these groups into categories, we come up with broad groupings of the main players in water management, as outlined in Table 2-5.

The public at large is not shown in the list, but it should be understood that each person plays a role.

These impacts illustrate how in TWM we are concerned not only with water management itself; we are also concerned with the impacts of land management on water (Figure 2-5).

Table 2-4. People and organizations impacting water resources

Category	Example of use or impact
Small water users	Well pumpers or water diverters for small businesses, groups of homes, or similar enterprises
Land and urban developers	Land development that alters the landscape and creates stormwater discharges and hydrologic modifications
Large corporate farmers	A water district or irrigation company that diverts or pumps water and discharges return flows
Small farmers	Small groundwater pumpers or dischargers from crop uses or livestock, including return flows
Water recreationists	Recreational groups for boating, fishing, or other recreation that involves water uses or impacts
Road departments	Highway and road transportation has large impacts on water through hydrologic modifications and nonpoint discharges
Nonpoint source dischargers	All nonpoint source activities, including agriculture, mining, urban runoff, waste sites, etc.

Table 2-5. Main players in water management

Category	Players
Water supply providers	Municipal, industrial, and agricultural water suppliers that divert water or pump from wells
Wastewater point dischargers	Municipal and industrial point dischargers
Stormwater and flood control agencies	All organizations that control storm drainage or flood control facilities
Land developers and road departments	Subdivision developers, road builders, road maintenance organizations
Farmers and resource extractors	Farmers and all classes of resource extractors have large impacts via the nonpoint sources they create
Instream flow groups	Groups with interest in instream flows, including lakes, for environmental or natural resources purposes
Navigation and energy organizations	Navigation and hydroelectric organizations (thermoelectric diversions are under "Water supply providers," above)

How the players create impacts on water systems

Now we know how impacts on natural water systems occur (storage, diversions, point discharges, nonpoint discharges, and hydrologic modifications) and who the main players are. In this section, we turn to how these players impact the systems and each other. Figure 2-6 illustrates how these impacts occur. The sources are shown on the outer ring and the impacts on the inner ring. The feedbacks and crossovers are shown with dashed lines.

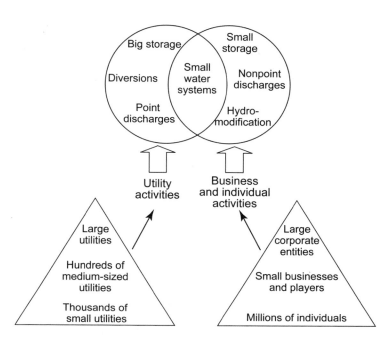

Figure 2-5. Water and related land management activities

Water storage

Water storage is created when dams back up stream flow to create reservoirs, or "buckets." The purposes of storage are to create reserve water supplies; to back up water for navigation, hydroelectric generation, and recreation; or to mitigate floods by managing the release hydrograph. Storage has beneficial effects to mitigate risk of drought water shortage, provide water for economic and social purposes, and reduce risk of flood damage. Instream flow interests, such as recreation organizations, could build storage, but they tend to rely on water released from storage owned by other entities because they usually lack the funds or authority or both to build storage themselves. An exception might be a state natural resources department that builds a lake to guarantee flow downstream.

One set of players in water storage comprises the owners of large dams, which are mostly the US Army Corps of Engineers, US Bureau of Reclamation, electric power companies, cities, irrigation companies, and industries. Another set is the group of owners of smaller reservoirs.

In the United States, about 78,000 dams are identified as large dams in the National Inventory of Dams. To receive this designation, dams must be more than 25 feet high, hold back more than 50 acre-feet or water, or be considered to present a significant hazard if they fail. State governments compiled the initial data and furnished it to the Corps of Engineers (2006). Considering there are about 3,000 counties in the United States, this averages to about 26 large dams per county across the nation. That number is substantial, and even dams that might look small fit into this category.

While some large dams are owned by the federal government, ownership is dispersed among many cities, districts, and even private entities. About 58 percent of US dams are privately owned. Local governments own about 16 percent and states about 4 percent. The federal government and public utilities own a smaller numbers of dams, but some of them are the largest in size. The primary purposes of dams in the United States (in order) are recreation, farming, flood control, irrigation, water supply, mine waste retention, and hydropower. State governments have regulatory responsibility for 95 percent of the approximately 78,000 dams within the National Inventory of Dams (ASDSO, 2004).

The companion feature to a dam is the reservoir that it impounds. Around the country, there are countless thousands of small lakes and ponds, compared to the some 78,000 reservoirs impounded by the large dams. As you can imagine, no inventory of these smaller lakes exists, but they are often created by small dams for watershed improvement, fishing, fish farming, parks and recreation, aesthetics, and wildlife management, and can even include gravel pits that remain after excavation.

Ponds can be so small that it hardly makes sense to think of them as water storage, but they are still considered as hydrologic modifications. Urban areas contain many such small ponds, some left over from old rural lakes, some developed for neighborhood recreation and aesthetics, some for stormwater control, some on golf courses, etc. Rural areas have fish ponds, small ponds for local water storage, and ponds that result from farm activities or wildlife management.

In urban stormwater management, a detention pond functions in the same way as a larger flood control reservoir, but it holds less water and responds faster. Most detention ponds located in urban areas have the goal of slowing runoff that has been accelerated by pavements built over natural lands. Stormwater detention ponds comprise significant hydrologic modifications, but they are intended to mitigate more severe hydrologic impacts of urbanization.

Reservoirs and ponds have effects on water quality and can themselves become water quality problems. One of the conditions that can creep up on lakes and reservoirs is the phenomenon of aging called eutrophication; when nutrients build up and other water quality changes occur, nuisance algae can grow and cause severe clogging problems of water quality decline.

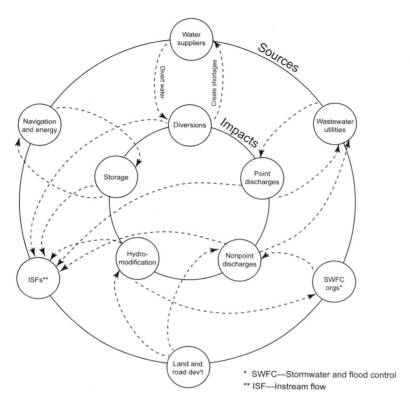

Figure 2-6. Sources and impacts of TWM actions

Diversions

Because we need water for many purposes, it is natural for utilities and other organizations to divert it from streams and pump it from wells. It is not practical to leave all water in streams without diversions. The important question is: what is a reasonable amount to divert so that balanced uses occur?

No one really knows how many water diversions there are in the United States, because no national database for them exists and no registration is required for diversions in some states. Diversions are made by water users for irrigation, thermoelectric power, public supply, industrial and commercial use, domestic use, aquaculture, mining, and livestock.

There are about 78,000 large dams in the United States

Some state governments have records of major water diversions. For example, the State of Colorado has tens of thousands of water rights of different types, and many of them have diversion points. These are maintained in the State Engineer's data files because the diversions have legal status under the appropriation doctrine of water law. Only a few states have records like this. Some states have policies to collect water-use data but have not provided the support for agencies to implement appropriate programs.

Although there is no national database of diversions, the United States Geological Survey (USGS) does maintain a water-use database that is based on estimation procedures (Hutson et al., 2004). The most recent USGS report shows that water use during 2000 in the United States was about 408 billion gallons per day (bgd). Prior to the 1980s, water use in the United States was increasing, but since 1985, total use has been mostly steady because the large withdrawals that occur for thermoelectric power and irrigation have been stable. Groundwater withdrawals during 2000 were 14 percent more than during 1985, but surface water withdrawals have been stable. Table 2-6, based on USGS reports, shows the year 2000 withdrawals for the United States.

Point source dischargers

In point source discharges, wastewater is collected and discharged at a single point. The main record of point source dischargers is of National Pollutant Discharge Elimination System (NPDES) permits, which are required for municipal and industrial (M&I) dischargers, M&I stormwater systems, and concentrated animal feeding operations (CAFOs). In the United States, about 500,000 entities are required to have NPDES permits. This includes several hundred thousand businesses in more than 50 industrial categories as well as about 16,000 municipal wastewater treatment plants (USEPA, 2001).

Table 2-6. Estimated water use in the United States in the year 2000 in million gallons per day (*mgd*)*

Type of use	Fresh water	Saline water	Total
Public supply	43,300		43,300
Domestic use	3,720		3,720
Commercial	Not reported		Not reported
Irrigation	137,000		137,000
Livestock	1,760		1,760
Aquaculture	3,700		3,700
Industrial	18,500	1,280	19,780
Mining	2,010	1,490	3,500
Thermoelectric	136,000	59,500	195,500
Total use	346,000	62,300	408,000

Source: Hutson et al., 2004.

* *Note:* The table reports water diversions only and does not include instream uses.

Point source dischargers are required to have permits that coordinate the strength of wastewater with the capacity of the receiving streams to assimilate the pollutants without violating stream standards. Water quality standards are set by considering designated uses, water quality criteria, and an antidegradation policy. Designated uses are set by government for the intended use of the water body. Water quality criteria are concentrations of pollutants and other measures of water quality. Antidegradation policies are rules to prevent deterioration of high quality waters (USEPA, 2006b).

> **There are some 500,000 point source dischargers in the United States**

Some people think that because point source controls are in place, the nation's water quality problems are solved. This is far from the case, however, because most of the degraded stream segments are caused by nonpoint sources and hydrologic modification.

Nonpoint sources

Nonpoint source (NPS) pollution is the leading remaining cause of water quality problems. It is caused by human activities on the land, and each person, as well as the three levels of government, has a role in stopping it. Most information about NPS pollution comes from the US Environmental Protection Agency (USEPA, 2006b). NPS pollution comes from many diffuse sources and is caused by runoff, which carries natural and human-made pollutants to lakes, rivers, wetlands, coastal waters, and groundwater. Categories of NPS are:

- Excess fertilizers, herbicides, and insecticides from agricultural and residential sources;
- Oil, grease, and toxic chemicals from urban runoff and energy production;
- Sediment from construction sites, crop- and forestlands, and eroding streambanks;
- Salt from irrigation and acid drainage from abandoned mines;
- Bacteria and nutrients from livestock, pet wastes, and faulty septic systems; and
- Atmospheric deposition and hydromodification.

Stream impairment from point and nonpoint sources

USEPA monitors water quality changes that occur from NPS and point sources through the reporting process specified by Section 305(d) of the Clean Water Act. USEPA (2000) collects and summarizes reports from state governments. It has changed its methods of reporting, so data

Table 2-7. USEPA findings on water quality

Total	Assessed (%)	Good	Good but threatened	Polluted
Rivers, total miles				
3,662,255	842,426 (23%)	463,441 (55%)	85,544 (10%)	291,264 (35%)
Lakes, acres				
41,593,748	17,390,370 (42%)	7,927,486 (46%)	1,565,175 (9%)	7,897,110 (45%)
Estuaries, sq. mi.				
90,465	28,687 (32%)	13,439 (47%)	2,766 (10%)	12,482 (44%)

Source: USEPA, 2000.

from the 1998 report is presented here. About 32 percent of US waters were assessed for this inventory. Reports were that about 40 percent of streams, lakes, and estuaries were not clean enough to support intended uses. Leading pollutants in impaired waters included siltation, bacteria, nutrients, and metals. Runoff from agricultural lands and urban areas were the primary sources of pollutants. Table 2-7 presents the results.

Table 2-8 summarizes the main sources of pollutants by water body type. For rivers and streams and for lakes, ponds, and reservoirs, the sources were agriculture, hydromodification, and urban runoff/storm sewers. Main pollutants were siltation, pathogens, nutrients, and metals. For estuaries the main sources were municipal point sources, urban runoff/storm sewers, and atmospheric deposition. An additional principal pollutant was organic enrichment and low dissolved oxygen.

The figures capture the largest water bodies, but not the many miles of small streams, the aquifers, and the many small ponds.

Table 2-8. Main pollutants

Rivers and Streams	Lakes, Ponds, and Reservoirs	Estuaries
Siltation	Nutrients	Pathogens (Bacteria)
Pathogens (Bacteria)	Metals	Organic Enrichment/Low Dissolved Oxygen
Nutrients	Siltation	Metals

Hydrologic modifications or alterations

Hydrologic modification is the least well defined category of impacts. USEPA's (2006b) explanation is that it comprises channelization and channel modification, dams, and streambank and shoreline erosion.

Dams were already included under "storage" above, so this categorization focuses on channels and their modification. Examples given earlier were:

- Stream constrictions for culverts
- Filling in of streams or channel changes for land development
- Stream maintenance activities
- Wetland disturbances
- Roads that block small drainages and divert water to ditches

This category can include small streams and ponds that often get clogged with algae or sediment.

USEPA's (2006b) explanation of hydromodification focuses on both water quality and quantity: "Hydromodification is one of the leading sources of impairment in streams, lakes, estuaries, aquifers, and other water bodies in the United States." It continues, "hydromodification activities . . . change a water body's physical structure as well as its natural function. These changes can cause problems such as changes in flow, increased sedimentation, higher water temperature, lower dissolved oxygen, degradation of aquatic habitat structure, loss of fish and other aquatic populations, and decreased water quality. It is important to properly manage hydromodification activities to reduce nonpoint source pollution in surface and ground water."

Road construction and drainage from the nation's some four million miles of roads is a large source of hydromodification. Pavement drainage is an NPS, but water diversion and changed quantities from pavement and water running in ditches are modifications. For example, a stream constriction for a culvert modifies the flow greatly and may cause erosion and sedimentation. New land development that causes filling in of streams and floodplains is a hydromodification. These can include channel changes, straightening, and relocation. Wetland disturbance is a hydromodification.

Hydromodification is a leading cause of impairment in US water bodies

Identifying hydromodifications

Given the vulnerability of headwaters, it is important to study small streams to assess the extent of hydromodification. USEPA's stream inventory for water quality reporting offers a way to map these small streams. USEPA (2005) recommends the use of the National Hydrography Dataset and mapping at 1:100,000 scale. However, states can use 1:24,000 mapping if they choose and report their use. Stream mileage is compiled from USGS mapping and USEPA's Reach File. In this file, reaches are defined as intervals of surface water between stream confluences or lake entrance and exit points. In 1982, the Reach File had about 68,000 reaches and 650,000 miles of stream. By the late 1980s, a new Reach File had some 170,000 reaches. This indicates that the Reach File had upwards of 2 million miles of stream, or more than half of the some 3.6 million miles of rivers reported in the 305(d) report. There are, however, many more miles of smaller streams that are not necessarily "significant" pieces of surface water for the USEPA definition but which might still be important for habitat.

Summary of players and impacts

This discussion reveals that the water environment is impacted by numerous large players and by multiple small systems and players. Figure 2-7 summarizes the physical cycle. The institutional aspects of making incentives and the regulatory system work to encourage better stewardship are explained in some detail in chapter 10.

The appendix to this chapter presents a case study in the form of a play that explains a typical scenario in which TWM is required to coordinate the players and control the impacts.

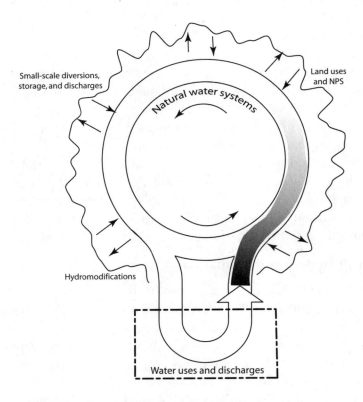

Figure 2-7. Water cycle uses, discharges, and effects

Summary points

- New water supplies are getting harder to find. Water demand patterns have changed, and in the future, water will cost more, and more conservation will occur. There will be a continuing reallocation of water from low-value to higher-value uses.
- The concept of water quality is very general, and no single index for it exists. There is disagreement about how water quality affects ecosystems and human health, and all stakeholder views must be included in standard-setting.
- Water quality problems are caused by how society lives and has lived, what society produces and consumes, how people and goods are transported, and how plans are laid for the future. In addition, new water quality concerns include pharmaceuticals in water, algal toxins, and distribution system water quality.
- With an aging population and more focus on healthcare, public concern about water quality is growing.
- The environmental needs for water can be measured by uses of soil systems, plants and biota, microbes, insects, and fish and wildlife.
- The Tragedy of the Commons is a slogan meaning that people take actions in their own interests and care for their own property but not for the common property of the public.
- Water management actions can be placed into five groups that include direct uses of water and land uses that affect water. These are water storage, diversions, point source discharges, nonpoint source discharges, and hydrologic modifications or alterations.
- Sustainable development must be a shared responsibility across society, and TWM offers a way to accomplish it through management actions that affect water as it passes through the hydrologic cycle.
- To be effective, TWM must include direct functions, such as diverting water and returning it to the stream, and indirect functions, such as land use, stream modification, and nonpoint source control.
- In addition to water managers, TWM requires participation by people and organizations that fall outside the world of organized utilities, industries, agencies, and regulators. Examples are small water users, land and urban developers, large corporate farmers, small farmers, water recreationists, road departments, and other nonpoint source dischargers.

Review questions

1. Describe the challenges of developing new water supplies, with emphasis on how these challenges are likely to unfold in the immediate future. What measures should water utilities take to prepare for them?

2. Explain the major issues in water quality today. Are point sources or nonpoint sources more difficult to control? Why? Explain the measures a water utility should take to inform its customers and stakeholders about water quality.

3. In TWM, the environmental needs for water must be considered along with those for humans. What are the major needs of water to be used by the environment?

4. The slogan Tragedy of the Commons is used widely to explain a vexing problem of water management. What is meant by this slogan and how does TWM respond to it?

5. Explain what actions are included in water storage, diversions, point source discharges, nonpoint source discharges, and hydrologic modifications or alterations.

6. Explain the roles in TWM of water managers and people and organizations that fall outside the world of organized utilities, industries, agencies, and regulators.

References

American Water Works Association. 1994. Principles of Total Water Management Outlined. *MainStream* 38 no. 11 (November): 4, 6.

American Water Works Association. 1996. Water:\Stats The Water Utility Database. 1996 Survey Complete Set. This product can be ordered from AWWA: http://www.awwa.org/Bookstore/productDetail.cfm?ItemNumber=53003.

Association of State Dam Safety Officials (ASDSO). 2004. *Dam Safety 101*. http://damsafety.org/. Accessed January 5, 2004.

Black, Peter E., and Brian Fisher. 2001. *Conservation of Water and Related Land Resources*. 3rd ed. Boca Raton, Fla.: CRC Press.

Hardin, G. 1968. The Tragedy of the Commons. *Science* 162: 1243–1248.

Hutson, Susan S., Nancy L. Barber, Joan F. Kenny, Kristin S. Linsey, Deborah S. Lumia, and Molly A. Maupin. 2004. *Estimated Use of Water in the United States in 2000*. USGS Circular 1268. Reston, Va.: US Geological Survey.

Singer, Phil, Cynthia Doherty, and L.D. McMullen. 2003. Panel on Drinking Water Quality. American Water Works Association Annual Conference and Exposition, Anaheim, Calif.

US Army Corps of Engineers. 2006. *National Inventory of Dams*. http://crunch.tec.army.mil/nid/webpages/nid.cfm. Accessed October 30, 2006.

US Environmental Protection Agency. 2000. *Water Quality Conditions in the United States: A Profile from the 1998 National Water Quality Inventory.* Report to Congress. EPA 841-F-00-006.

USEPA. 2006a. *Introduction to the Clean Water Act.* http://www.epa.gov/watertrain/cwa/cwa4.htm. Accessed November 28, 2006.

USEPA. 2006b. National Management Measures to Control Nonpoint Source Pollution from Hydromodification. Draft. http://www.epa.gov/owow/nps/hydromod/index.htm#01. Accessed October 30, 2006.

USEPA, Office of Water. 2001. *Protecting the Nation's Waters Through Effective NPDES Permits. A Strategic Plan, 2001 and Beyond.* EPA-833-R-01-001. Washington, D.C.: US Environmental Protection Agency.

USEPA, Office of Water. 2005. *Guidance for 2006 Assessment: Listing and Reporting Requirements Pursuant to Sections 303(d), 305(b) and 314 of the Clean Water Act.* July. http://www.epa.gov/owow/tmdl/2006IRG/report/2006irg-report.pdf.

Water Quality 2000. 1992. *A National Water Agenda for the 21st Century.* Alexandria, Va.: Water Environment Federation.

Wikipedia. 2007. *Tragedy of the Commons.* http://en.wikipedia.org/wiki/Tragedy_of_the_commons. Acccessed May 18, 2007.

—Case Study—

Maintaining Supply While Preserving The Resource

Illustrating TWM Through a Play

Why a play?

Water management is so complex and involves so many different issues and players that it is impossible to reduce it to a set of formulas. A better way to illustrate the action is through case studies, which are very similar to stories. The pages that follow present a short play about events in a fictitious watershed. In the nation's some 3,000 watersheds, each about the size of an average county, many dramas are played out in managing water. This fictitious account illustrates how things might play out in one of them. The play illustrates a range of issues that occur repeatedly in TWM scenarios.

Setting

The setting is in a region in the northeastern United States, with three cities, located in two adjacent counties, astride the Vienne River, which becomes an estuary and discharges into the Atlantic Ocean (Figure 2A-1).

The region is experiencing moderate to strong growth, as well as demand for second homes. The cities and the rural areas are adding new subdivisions and suburban homes, hobby farmers, strip malls, and commercial, light industrial, and other typical developments. On the Vienne River, Lake Bourchet provides recreation and a source of water supply

Characters (in order of appearance):

James Roberts
　President, Toqueville City Council
Fred Jones
　Utility director, Toqueville
Jane Cleary
　Associated Farmers and Fishers
Elizabeth Lode
　Secretary, State Department of the Environment (SDOE)
Bill Ouellette
　Governor
Fred Ross
　Attorney, head of SDOE enforcement staff
Lex Hughes
　Director of Water Quality, SDOE
David Klim
　Director of Water Resources, SDOE
John Stevens
　Director of Parks, Recreation, and Wildlife, SDOE
Judy Jagger
　Special assistant to Secretary Lode
Micah Alexander
　State USGS director
Galvin Schmidt
　Professor at local university
Steve Geld
　Finance Director, SDOE
Jim Summers
　Chief scientist, Environmental Studies Bureau, SDOE
Linda McBride
　Staff member, City of Beauvais
Jane Jacobs
　Lead Scientist, Water Aquarion Consultants

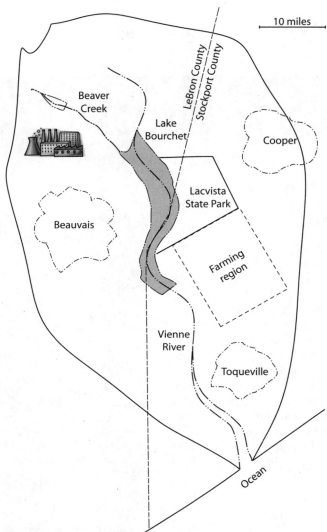

Figure 2A-1. Vienne River and estuary

for the City of Toqueville. The cities of Cooper and Beauvais draw water from springs and tributaries and discharge sewage into Lake Bourchet.

Issues

Water quality has deteriorated in the basin and in the river, and there a general sense of tension about how to solve the problems. Lake Bourchet is losing its clarity and tourist appeal at the same time that pressures to use more water and to discharge more wastewater into tributary streams are increasing. In striving to meet the needs of the developing area, the water utilities and wastewater utilities need to expand and renew their treatment plants and infrastructures. Because of the water quality issues, the cities must spend more money to meet regulations. The industries would like to be able to expand, and the electric utility would like to build a new steam plant in the area but is worried that cooling water won't be available. Environmental groups have noticed the trend toward lower water quality and loss of habitat and think that regulations are not being enforced.

Opening scene— City Council, Toqueville, October 11

James Roberts, president of city council: "Meeting will come to order. On the agenda tonight, we have a presentation by our utility director, Fred Jones, about a possible rate increase, and Ms. Jane Cleary will represent the Associated Farmers and Fishers to explain an issue they have. Mr. Jones."

Fred Jones, director of water and wastewater utility: "Thank you, President Roberts. I am pleased to appear before the council on this important matter. As you know from previous briefings, the combination of growth, meeting permit requirements, and renewing our aging infrastructure has put us under a lot of stress. We have a lot of problems from increased treatment costs, and we need to expand our river pump station and replace worn-out parts on our wastewater treatment plants and facilities. I am here with a rate study that shows we need 7 percent per year for the next five years to pay for improvements. With our current utility budget of about $20 million per year, that'll give us $1.4 million in new funds the first year, and by the fifth year we will be getting about $8 million per year more than today. That will allow us to float a bond issue, build the facilities we need, and stay out of jail, hopefully."

Roberts: "Wow, that's a lot of money; what will the ratepayers think?"

Jones: "I don't know, but we really don't have any choice, because water quality in our raw water is way down, we're getting pressure from the regulators for wastewater, and the system can't take any more new customers."

Roberts: "Well, Mr. Jones, we will hold a study session on this and see what happens. Thank you."

Roberts: "Ms. Cleary, you're here to represent a group in the county?"

Jane Cleary, representative Associated Farmers and Fishers: "Mr. President, thank you for letting us have time on your busy agenda. To cut to the chase, our group of farmers and fishers is strapped. It's hard to make a living in farming today, and most of us are only able to make it because we can supplement our farm income with fishing. Now, the quality of the river water has been getting worse and worse, and the fish are dying off. Some have fish disease and are not fit to eat. Frankly, we don't know what we're going to do, and we need some relief. If the town expands its pump station and takes more water from the river and doesn't fix its sewer plant, we are going to be out of business soon."

Roberts: "We've heard of your issues, but we need a better understanding of them.

Mr. Jones, you know about the river. What's the story?"

Jones: "Well, it's true the river is getting worse. You know that upstream you've got two growing towns in Beauvais and Cooper, a big fertilizer plant, and a lake with a lot of boats coming in from all over. We think it's a combination of things, and I am not sure what to do about it. One part of our rate increase request is aimed at better water quality, but we can only control our part of it, and I don't think that's going to help Ms. Cleary's group."

Roberts: "Surely we can do something. I'll go see the governor. He is a friend of ours and has more resources than we do."

Governor's office, October 25

Governor Bill Ouellette: "What's this about?"

Roberts: "Governor, we have a problem. As you see from our briefing book, our region has three towns, two counties, and a lot of other people involved. We are not sure what to do. Our river seems to be going downhill, we have some economic problems with rate increases and aging infrastructure, and your regulators are breathing right down our necks."

Governor (turning to his environment secretary): "Well, Liddy, it sounds like another issue for the Environment Department! What's the story?"

Elizabeth Lode, secretary of environment: "Governor, the river is meeting standards, but there are some strange things going on, particularly with algae blooms and fish disease. Might be something we just don't understand."

Governor: "Do we have some studies?"

Lode: "We have our own monitoring data, but the only overall study is from a professor at the state university, and it's inconclusive. He thinks there might be too many nutrients in the water. We are not sure if he's right or, if he is, where they come from. Some say it's the city sewage effluent, some say it's the farm runoff, and some say its the factory up there."

Governor: "Well, we can't do anything if we don't know who's causing the problem. Let's have a meeting up there in the county and see who knows what. Liddy, why don't you see what you can set up?"

Department of Environment, October 26

Secretary Lode calls the division heads together to brainstorm on a strategy. In attendance: Lode; Lex Hughes, Water Quality; David Klim, Water Resources; John Stevens, Parks, Recreation, and Wildlife; Fred Ross, attorney; Judy Jagger, special assistant to the secretary.

Lode: "Well, each of you were briefed; what do we do? Should we have a public hearing in the area to gather some testimony?"

Hughes: "Madam Secretary, we have already studied that situation, and they are overreacting. All the waters meet standards, and there really isn't anything we can do."

Klim: "Our people have reached the same conclusion. We are on the right track, but we have to spend more money on treatment and infrastructure, and it's up to them to come up with it. They are probably looking for a handout from the state."

Stevens: "Wait a minute. That area has been going downhill for a long time, and there are plenty of violations that your people have missed. That fertilizer plant up there is spewing nitrogen out its stacks, and the rain is dropping it into the watershed. It looks like a green jungle in places. The wastewater plants are not working well, and you can tell by the odor. We've had several algae blooms in the lake, and the fish are looking poor all the way to the ocean. Why can't something be done?"

Lode: "Hughes, what about that?"

Hughes: "Some of that is true and some is overblown. It sounds like your people are singing the tune of the environmentalists, and they always exaggerate things. There's nothing out of the ordinary in the basin, and, after all, we can't tighten the regulations. The cities and businesses are already on the ropes from that competition from Asia."

Lode: "Well, I can see that you aren't in agreement. We need a fresh look at the situation. Governor Ouellette wants some answers within 30 days. Judy, pull together a task force and get me a working paper on this by two weeks."

Department of Environment, two weeks later, November 9

Judy Jagger: "Madam Secretary, here is the result of your inquiry about the Lake Bourchet region. We found out that there is a lot of data but not too many conclusive results. It's hard to figure out what the main issue is. We'll have to do some research before we can recommend specific actions."

Lode: "More studies! The governor wants action now! Isn't there anything we can do?"

Jagger: "We think it would be good to have a public hearing in the area and see what other people think. After all, the people who came to see the governor are from the lower basin, and we really don't have any input from Cooper or Beauvais or the industries."

Lode: "Sounds like it would be more of a regional workshop of water people than a public hearing. After all, the public would just get confused, wouldn't they?"

Hughes: "We have some monitoring work going on, and a regular group meets from time to time. We could pull in some more players and make it a round robin to talk about what everybody's doing; then we would have a better idea of the big picture. Want us to organize it?"

Lode: "That sounds like a good plan for now. After we see what everyone is doing, we will have a better idea of what to do. See if you can do it within two weeks, please."

Hughes: "Will do. I'll work with Klim to make sure we include all the stakeholders."

Regional meeting, December 3

Director Hughes confers with Director Klim, and they set up a working meeting of the players. Secretary Lode attends to welcome the participants. The sign-in list shows 38 people as follows:

Federal regulator	1
Water supply directors from cities	3
Wastewater directors	3
Stormwater utility director	1
Corps of Engineers engineer	1
USGS district chief	1
State water resources director	1
State USEPA director representative	1
Electric power generator representative	1
Land developer	1
Large corporate farmers	2
Groundwater pumper	1
Environmentalists	3
Bass Fishing Association representatives	2
Consulting engineers	3
Water and environmental attorney	1
Think tank representative	1
Association executive representing vendors	1
Department of Natural Resources representative	1
County representatives	2
Industry representatives	2
Secretary Lode and staff	5
Total	38

Lode: "Welcome. I am impressed by how many professionals have made a commitment to attend this important meeting. We will begin with a short presentation by Director Klim, then take your responses."

Klim: "The basic issues that triggered this meeting revolve around water quality problems in the lower Vienne River as reported by Jane Cleary of the Associated Farmers and Fishers at a meeting of the Toqueville Council on October 11. This was corroborated by Fred Jones to the group, then reported to Governor Ouellette at a meeting on October 25. Governor Ouellette asked Secretary Lode to look into it, and she commissioned this paper I will present.

"The report by the Associated Farmers and Fishers is about the deteriorated quality of the river water, which they say has been getting worse and worse, with fish disease and kills. They are also worried about Toqueville's plans to take more water out of the river and pump more sewage into it.

"Our initial inquiry showed that river water quality is getting worse, but it still meets standards. Upstream are two growing towns, a fertilizer plant, a lake that is under pressure, and a lot of individual development activity, septic tanks, and the like. It's a complicated situation.

"We looked at the data sources and found that there is quite a bit of data but not much information and not much wisdom, for sure. In other words, we see some pieces of the puzzle but not how it fits together or what the picture will look like after it comes together.

"The environmental data we have comprises USGS water quality and stream gauge data. It gives a picture of the basic parameters, but there is a lot of variation, and it is not clear what the trends are. We also have diversion and discharge data for the towns and industries, and we have data on the types of farming and recreation activities going on. No one, however, has put the pieces together. The closest thing to an integrated report is by Professor Galvin Schmidt of the university, who assigned the problem to a group of students as a class project. I've asked Professor Schmidt to present their report so that we can get an overview."

Professor Schmidt: "Thank you, Director Klim. My class is a master's level course in watershed hydrology, and we look at all aspects of the hydrologic cycle, including water quality. I had the students use the Vienne River as a case study and look at the natural flows, diversions, outflows, discharges, and water quality constituents. They interviewed a number of you, visited the sites, took photos, and compiled their findings into the report you see here.

"Please understand that this represents an unfunded effort of a few weeks. The major value of the report is the collection of diverse data and the attempt to integrate it. The students found that, although the river meets standards, there are no standards for algae. In fact, no one understands what causes the algae blooms that have clearly occurred. There are some nutrient data, but the link between these data and the algae blooms is not explained.

"By the same token, we lack information about the links between algae, other water quality constituents, and fish health. The stocks of anadromous species, such as salmon and herring, have declined, but other species have increased. Natural cycles are at work as well as those affected by water quality.

"There you have it. We found that water quality standards appear to be met, but it is not clear if those standards are appropriate. We found that river quality levels are declining but do not know exactly why. We recommended an extensive research program to uncover some of the secrets of the river and see if it can be made healthier."

Klim: "Thank you, Professor Schmidt, and thanks to your students. Although your team did not collect original data or reach definitive conclusions, you gave us valuable insight. Now, let's hear from USGS."

USGS director Micah Alexander: "We monitor river water quality and diversions as part of our ongoing national program, but our program is limited to the constituents necessary for the required reports. We think that a lot more data is required to provide an integrated picture of the river's water quality, and the data has to be analyzed to create a full assessment. However, we are not funded to do that, and apparently no one else has the responsibility. We are glad to make our data available to the group."

Lode: "Lex, how does it look to the water quality agency?"

Lex Hughes, water quality director: "Madam Secretary, that's about all the integrated information we have. We have data that enables us to see if standards are being met, but there are questions about the standards. Clearly, the gross pollution of yesteryear is gone from the river, but we are getting algae blooms and seeing more fish kills. Our job is to monitor, regulate, and report. We have reported that the river meets all the standards, but we noted in our last report that the algae problem just showed up and is unexplained. The big picture is hard to assess, and we are not able to clamp down on any parties unless we can verify their role in the problems."

Lode: "Well, I can report back to the governor, but we need some daylight here. There isn't a clear villain, is there? We'll figure something out."

Governor's Office, December 10

Meeting between Governor Ouellette, Secretary Lode, and Councilman Roberts.

Governor: "Well, Liddy, what did you learn?"

Lode: "Governor, I know you like to take action, but we don't know who is causing these problems. We need some more time to study it. I commissioned a study out of my office and then we had a meeting of the agencies and professionals in the area. There are data but no one really knows what's going on."

Roberts (looking at Lode): "Madam Secretary, we already knew that. It took you a couple of months to come up with that nonconclusion?"

Governor: "Look, James, I know you want action too, but we can't afford to go off half cocked. We've got to be sure that if we clamp down on somebody we'll get the results we need. After all, you have been one of the loudest voices against government interference in private or local affairs. I tell you what. If you'll put some of your people on a panel Liddy will appoint, we'll conduct an inquiry and figure out the best thing to do."

Roberts: "Well, we already know that it's those people upstream, but we'll play along. I'll have our city manager put some resources into it."

Governor: "Good. Liddy, let's get a panel of experts together and have them report to an oversight group that represents the watershed. We ought to be able to do it in six months, what do you say?"

Lode: "We can do it, Governor. I'll send you a report in about two weeks, and we'll get it going."

Governor: "Sounds good. If you need my help for any part of it, let me know."

(Councilman Roberts leaves.)

Governor: "Liddy, this is just the tip of the iceberg on this issue. I've heard about it through other channels, and I'm counting on you to get positive traction and to find some daylight in a situation that could get nasty and threaten my legacy as an environmental governor. You've got a big stake in it, too."

Lode: "Bill, you and I see it exactly the same way. I think I see some avenues to make this a lot better, and I'm optimistic. I'm going to need some money to hire a consultant and will have to lean on you for some of it."

Governor: "As long as it isn't too much, just talk to my assistant and we'll find the money. This is high priority in my administration."

Department of Environment, December 11

Lode meets with division chiefs, enforcement staff, finance director, and special assistant.

Lode: "Well, Judy and David, you've had a chance to think about it. How should we put this group together?"

Klim: "Madam Secretary, we have talked about it in the Department of Water Quality, and we think we have a good handle on part of the issue, but we lack some data, and we don't have a watershed model. We think a budget is needed to hire a consultant to model the watershed and to collect a few pieces of missing data. We are also going to need some lab analysis work, particularly on the biological side. We think that about $150,000 for the six-month effort will be a minimum amount, and we can get the rest from contributed services."

Lode: "That's a lot of money, and we don't have it in the budget, unless we can cobble it together from federal planning funds. (Turns to finance director Steve Geld.) Steve, can you find your money?"

Geld: "Madam Secretary, we could find part of the funds from our water quality accounts, but we need about half of it from outside the department. It makes sense that part of it should come from economic development planning funding, but we'd have to get the governor to direct those funds to us."

Secretary Lode talks to the governor's special assistant and arranges for the funding. Klim and Jagger meet and work out the lineup for a study team, work with the secretary's office to identify citizens in the basin for the oversight group, and set up the first meeting of the basin study group.

Lodge at Lacvista State Park, January 11

Department of Environment hosts the study team and oversight group. Meeting is also attended by basin stakeholders, making a total of about 100 people.

Lode: "I want to thank you all for coming today to this important meeting. We have here a historic opportunity to reverse the gradual slide in environmental conditions that has hurt the economy and the people in this area. This is a classic environmental problem. Clearly, something is out of balance, but we are not sure of what it is. We need to use the valuable water resources here for positive purposes, but we can't stress them too much or they will fail. In other words, we are looking at a challenge to sustainable development.

"As you know, Governor Ouellette was elected on a platform that stressed a balance between economic development and environmental protection, and he asked me to assure you that this administration is 100 percent committed to finding a solution to this issue, no matter how long it takes. We will stick with you and find a solution that works for everyone.

"Now, I've asked Jim Summers to lead the study team, and he will introduce the team to you. As you know, Jim joined the department two years ago as our chief scientist. He completed a tour as USEPA's regional water quality manager, and he holds degrees in environmental science and biology. His combination of a science and management background equips him to lead this study. Jim."

Summers: "Thank you, Madam Secretary. We are using a management model developed by USEPA and its partners based on the National Estuary Program, which was established under the Water Quality Act of 1987. While our problems here involve the estuary and the upstream reaches, this model seems to offer what we need.

"As you know, estuaries around the world are threatened. Only a fraction of them are even monitored for the most damaging pollutants. Pressure usually comes from population growth, agriculture, industrialization, and fisheries. Typical problems include nutrient enrichment, eutrophication, and nuisance algae. These cause low oxygen levels, loss of submerged vegetation, and fish disease and species changes, as well as other problems. Fortunately, we are not dealing with a lot of industry here, and it does not appear that our problems are as bad as they could be.

"In creating the National Estuary Program, Congress realized that the main challenge was to organize successful cooperative efforts, but these are necessary to develop overall estuary recovery and protection programs, with each party taking a role.

"USEPA looked at what it had learned during the Chesapeake Bay and Great Lakes studies, and created a four-phase model management approach to balance conflicting uses: planning and a management framework, characterization and problem definition, a comprehensive plan, and implementation. These require the governor's nomination, the convening of a management conference by USEPA, estuary characterization, a comprehensive plan, implementation plans, and continuous monitoring.

"The problem on the Vienne River is not as large in scope as some, but it has all the elements that you find in these estuary programs. In fact, these are what they call in the trade 'shared governance' problems because they require cooperation in identifying the problem and solving it.

"That's the overview. Now, what we've put together here is an oversight committee with elected and appointed officials, environmental managers, local scientists, and citizen representatives from interest groups. The study team is 'lean and mean,' with representatives from the Department of the Environment, USGS, and the three major cities in the basin. We have already met and created a draft request for proposals for the consulting study, and I'll ask Ms. Linda McBride from Beauvais to present that to you now."

McBride presents a RFP that specifies a four-month model study, data collection, public meetings, and a report.

Summers' Office: Selection of consultant, February 21, Notice to proceed on March 21

Summers (to selection committee): "Well, we got three proposals. We have to follow department procurement rules. What do you think?"

Geld: "I think that we ought to rate them in terms of the advertised criteria, which are expertise in modeling a water system, staff availability, and demonstrated past performance."

Summers (noting agreement): "Well, we're prepared to do that. (Summers goes to whiteboard and writes criteria across the top and three firm names along the side). OK, if each of you will enter your scores from 1, best, to 5, worst, for each firm and each criterion, we'll see if we have a meeting of the minds."

This exercise leads to a clear preference for Water Aquarion Consultants, with Jane Jacobs as lead scientist.

Consultant's team meeting with study committee, April 15

Consultants Jacobs meets with study team, presents PowerPoint.

Jacobs: "Thanks for meeting with us today. I want to show you the results so far and see where we should go next with this. These are the results of the water quality model, where we look at ten constituents from the headwaters through the estuary. As you see, we calibrated the model with the data we had, and it works fairly well. The problem is, we have little nutrient data and no data to speak of on the algae problem. In other words, we can look at the standard parameters of oxygen, temperature, and so on, but I am not sure we are getting to the root of the problem. We sent some fish tissue in for analysis, but it was inconclusive, and our studies of the algae show that it is the blue-green algae species *Anabaena*, but that may or may not be useful information. There are no patterns here to pinpoint one source or another. What do you think we ought to do next?"

Discussion ensues among study committee members.

Summers: "Jane, we see the problem. There are so many variables that we can't study them all or nail everything down. For your report due in a month, we'd like you to integrate the data and give us your best conclusions, given what you have."

Study Committee meeting and Consultant's interim report, May 15

Jacobs: "We looked at all of the data and find that the dischargers are generally in compliance, but we are not sure the water quality standards are adequate, and we don't have a handle on the nonpoint sources. Our overall recommendation is to recognize that you have a system in stress and you are going to have to do an in-depth monitoring and modeling study to find out the critical variables."

Summers: "Well, folks, we didn't think it would be easy. We clearly have a complex situation here but no smoking gun. We can't just go back to the governor and the secretary and tell them that more study is needed. We just did the 'more study.' We think some action is needed one way or another. We are going to recommend a regional action program with three basic elements: one will address municipal and industrial wastewater discharges, one will address nonpoint sources, and the third will address continued study and monitoring."

Briefing for towns, industries, counties, and other leaders, May 22

Secretary Lode: "As you know, we've been committed to working with you to find a solution to the water problems here. I wish I could bring you today a final solution that will work quickly, cheaply, and well. However, I can't. We've completed a $150,000 study, used the best data we could find, and employed the best minds. They have concluded that we don't know exactly what the problem is. That's the bad news.

"The good news is that we now have a regional group that is cooperating with each other to find a solution to a problem we share—cleaning up the Vienne

River for everyone. Therefore, I am announcing today a three-point program where we will work in concert over the next three years toward a big improvement in the river. Here are the program elements:

"First, we are asking the state water pollution commission to review the nutrient standards for the river and to consider more stringent limits. Then, all wastewater dischargers will have their permits reviewed and will be required to curtail discharge of nutrients to meet any new, more stringent standards. We recognize that this will be costly, and we are making low-interest loans available from the state revolving fund.

"Next, we are forming a Vienne River Watershed Association with the mission to assess and improve water quality in the basin using all measures available. I will be the interim chair, and we are inviting all of you to join.

"Finally, my office will fund a continuing monitoring and modeling program to review all sources and remediation measures toward measuring improvement and trends so that we can adjust these regulations as needed."

Meeting in Governor's office June 6

Lode: "Well, Governor, we did the best we could. If the data isn't there and you don't know exactly what to do, you can't shoot in the dark. There are too many people involved and too much money at stake. We have launched a program that seems sure to work, but there will be some unhappy people, and it's going to cost some money. We've already heard about resistance to the increased spending from all three towns, and the electric utility has sent their lawyers in to investigate the procedures we used. Moreover, the farmers are starting to chafe at the idea of nonpoint source controls. It seems like everyone is just a little bit mad. Maybe that shows we got it right."

Governor: "Liddy, you did a good job, and Jim Summers did, too. I haven't heard any real negatives, and my people tell me that folks in the basin like seeing the movement. We'll have to stick with this and see it through. Only problem is, you and me will be gone in three years. Who is going to follow up?"

Lode: "We think the new Vienne River Watershed Association will have to hold everyone's feet to the fire. Someone has to. I don't think we can count on state government to do it."

Governor: "I think you're right. Now, let's get to work and figure out who is going to be on that commission."

End of play and lessons learned

Water sagas often end this way, with more studies, calls for long-term solutions, and questions about how to sustain the action.[6] It is an unusual situation when a definite conclusion to a problem like this can be identified.

In this case, we have a Tragedy of the Commons type problem, but no clear cause is evident. It is difficult to develop a clear scientific cause for many water problems, and at the end of the day you have to do something, even when you are not sure you have it exactly right. This is one reason that people say that water problems are political.

---

6 This case study is actually closely related to one on the East Coast that unfolded a few decades ago and resulted in the same general outcomes. I have checked on the status of the water and have been told in general, "Well, we've not been hearing about problems there recently." That's a success story. There's no finish line or celebration party, but you know that the outcome is an improvement in the situation.

You see, if you are one of the utilities or dischargers who must spend more, and if no one is sure that the expenditure will help all that much, is it the right thing to do? If you take no action, the problem is sure to get worse. If you compel people to spend, and they choose to resist, you have a possible court battle. This is where TWM can pay big dividends, by encouraging the players to work out a solution that wins for all.

Notice in the story that there is not a logical group to pull things together. The convening authority actually came out of the governor's office. Otherwise, no one would step forward to take responsibility to integrate the views. Notice also that the only integrated study available was a low-budget ad hoc effort by a university professor, and this was just happenstance. There is also a lot of the "it's not my problem" syndrome evident in the story. Data to confirm or deny theories about the causes of problems is lacking, and people always can say, "Let's collect more data." Notice also that no one has responsibility for nonpoint sources and the hydrologic modifications. It was necessary to organize a watershed association to even get started on those. The cities can continue to gain access to water and to discharge permits for wastewater, but existing standards may tend to lower water quality and not deal with issues such as the clarity of the lake. Moreover, some of the lake problems may be due to overuse for recreation and not be under jurisdiction of the water quality authorities. These points were not discussed in the story, but they are evident from the setting.

Another observation from the play: The roles and points of view are predictable. An environmentalist can say, "We need some action" (this time it was the farmers and fishers). The state USEPA director can say, "We have a study underway, but it won't be completed for six months because we have to do more monitoring." A water recreation representative or environmentalist could say, "Another study? We need action now!" The convening authority can say, "Well, it looks like we shared some ideas; let's have another meeting in about a month to see if anything new develops." These are typical positions and points of view, but some forcing functions are needed to move the situation off of dead center sometimes.

CHAPTER 3

TOTAL WATER MANAGEMENT: VISION, PRINCIPLES, AND EXAMPLES

We have learned that TWM is a framework for principles and practices that lead to sustainable use of water resources and the "exercise of stewardship of water resources for the greatest good of society and the environment," but how does it work? This chapter explains its principles, processes, and practices. It includes a detailed definition and compares TWM to related concepts such as Integrated Water Resources Management and the European Water Framework Directive.

As with other visionary concepts, knowledge about TWM is heavy on principles and light on proven practices that work. So the problem is not so much to develop the concepts as it is to make them work. In other words, the challenge is to "walk the talk." The obstacles are introduced in this chapter and explained in more detail in chapter 10.

Fundamental concepts and definitions of TWM

A well-known book by Robert Fulghum is titled *All I Really Need to Know I Learned in Kindergarten,* and that idea is partly true with TWM. You can define TWM simply or in great detail. The simple explanation gets you most of the payoffs. If you define it in great detail, you get lost in the complexities. We need a simple way to express what we are talking about.

Water leaders have been looking for this simple-sounding concept for a long time, and an AWWA committee originated the TWM concept in the early 1990s. Appendix A presents AWWA and AwwaRF documents and concepts that trace the development of the TWM concept. These include a 1994 White Paper that outlines TWM's ideas in detail and a formal definition from an AwwaRF (1996) workshop. Also included are AWWA's policy statement on "Developing and Managing Water Resources," which contains TWM ideas and a definition from AWWA's *The Drinking Water Dictionary*.

This chapter builds on AwwaRF's definition to fully explain TWM and to provide examples to illustrate the concepts. Examples are necessary so that the explanations will not be too abstract.

Let's start by comparing AwwaRF's definition with the definition in *The Drinking Water Dictionary*. As you see from the two statements in Table 3-1, similar ideas can be explained in very different ways. In this case, one definition (AwwaRF's) emphasizes stewardship and the other (Symons, Bradley, and Cleveland, 2000) emphasizes management.

Table 3-1. Two definitions of TWM

AwwaRF definition of TWM	*The Drinking Water Dictionary* definition of TWM
Stewardship of water resources for the greatest good of society and the environment.	Management of water resources with a comprehensive approach to balancing resources, demands, and environmental issues.

Sources: AwwaRF, 1996; Symons, Bradley, and Cleveland, 2000.

Despite this difference, the ideas converge in the sense that stewardship means to represent the owner of water in the management of water. Since water belongs to everyone, managing it well is stewardship of our common property.

Of course, there is much more to TWM than these brief statements. TWM is a paradigm for managing water for maximum benefit to society and the environment and a conceptual framework for its principles, processes, and practices. A *paradigm* is a pattern or model of how things are done (Wikipedia, 2006), and a *framework* is a structure on which to hang a set of principles and practices. A *principle* is a rule for action and a *process* is a collection of actions that leads to a result. Figure 3-1 illustrates how these relate to each other. Practices, not illustrated in the diagram, are habitual ways of performing tasks and processes.

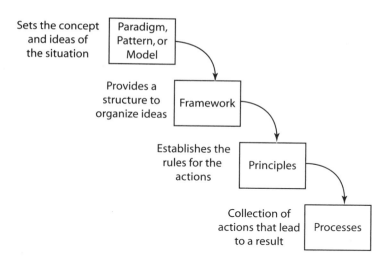

Figure 3-1. How a framework organizes principles and processes

What is water management?

Backing up a step, we ask: If TWM is "total" water management, what is "water management?" Even that simple-sounding phrase elicits controversy. Back in the 1970s, the American Society of Civil Engineers (ASCE) created a technical division and journal about it, the *Journal of Water Resources Planning and Management.*

Water management is often termed *water resources management* to emphasize that the *resource* is being managed, as opposed to the more limited concept of moving the water itself around, as in a pipe system. Scientists struggle to define it, because anytime you collect general words in a phrase, the phrase can be defined in different ways by different people.

If you tell someone you work on water resources management, they may or may not know what you mean. A problem of clear and consistent language arises because the words are so abstract. Say each word in the phrase "water resources management" can have three different meanings. Then, the phrase can have 3^3 meanings, or a total of 27 interpretations. Note that even "water" has different meanings, such as seawater, drinking water, raw water, piped water, and so forth.

After considering the many definitions of water resources management and in thinking about a field that includes a number of water sectors, I formulated this simple definition, which is not unlike the TWM definitions explained above: "Water resources management is the application of structural and nonstructural measures to control natural and man-made water resources systems for beneficial human and environmental purposes (Grigg, 1996)." A shorter version is: "Water resources management applies management tools to control water infrastructure systems for beneficial human and environmental purposes."

> **Water resources management applies management tools to control water infrastructure systems for beneficial human and environmental purposes**

TWM is a higher-level concept than water resources management, which emphasizes the control of water more narrowly. The emphasis of TWM is on management with stewardship, which draws in the larger society to work with water resources managers.

However you define TWM, its aim is to manage water resources in a "total" way. The important question is, how can TWM have practical value when it involves so many big-picture issues? The answer must be to show *how* structural and nonstructural measures *can be applied* to control water resources systems to result in *beneficial human and environmental purposes*. In other words, to implement TWM we confront the old saying, "the devil is in the details."

AwwaRF's detailed definition of TWM

The details of TWM were worked out by a group of more than 30 water industry professionals at the 1996 AwwaRF workshop. Here is the definition the group developed after two days of intensive work:

> Total Water Management is the exercise of stewardship of water resources for the greatest good of society and the environment. A basic principle of Total Water Management is that the supply is renewable, but limited, and should be managed on a sustainable use basis. Taking into consideration local and regional variations, Total Water Management encourages planning and management on a natural water systems basis through a dynamic process that adapts to changing conditions; balances competing uses of water through efficient allocation that addresses social values, cost-effectiveness, and environmental benefits and costs; requires the participation of all units of government and stakeholders in decision-making through a process of coordination and conflict resolution; promotes water conservation, reuse, source protection, and supply development to enhance water quality and quantity; and fosters public health, safety, and community goodwill.

This definition will be used throughout the book to explain concepts of TWM. While it exhibits good thinking, other definitions can be developed, such as the one for Integrated Water Resources Management.

TWM compared to IWRM and similar concepts

Of the concepts related to TWM, perhaps the one with most traction is Integrated Water Resources Management, or IWRM. IWRM is popular with intellectuals and with some practitioners, although people who use it do not always mean the same thing. Figure 3-2 illustrates how IWRM is meant to integrate policy sectors and water management purposes.

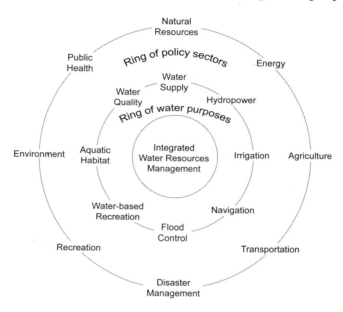

Figure 3-2. IWRM policy sectors and purposes

As with other intellectual frameworks, it is hard to identify the origins of IWRM, but water leader Gilbert White (1998) traced its history back more than 50 years. My studies trace its ideas back to at least 1917, when the Flood Control Act called for "a comprehensive study of the watershed." Figure 3-3 shows how IWRM evolved.

Pre-1900	1950s and 1960s	1970–2000	After 2000
Stage 1	Stage 2	Stage 3	Stage 4
Poor water management	Postwar development	Problem emergence,	Congested world
Urban sanitation poor	Growth stresses water	analysis, and recognition	Environment/society
Little environmentalism	Environmental crisis hits	Search for TWM paradigm	need TWM/IWRM

Figure 3-3. Development of TWM/IWRM concepts

The Global Water Partnership (2004), which is a leading sponsor of the IWRM concept, has developed a toolbox of methods for IWRM. Information available in the toolbox includes policy guidance; operational tools; case studies; and references, organizations, and websites. The structure for the tools is said to be based on three fundamental elements of IWRM:

- The enabling environment or rules of the game created by legislation, policy, and financing structures;
- Institutional roles of the players, with capacity-building supporting their functions; and
- Management instruments for water resources assessment, demand management, public information and education, conflict resolution, regulatory devices, economic measures, and information and communications.

The abstract language of IWRM leads to long and complicated explanations, such as one presented by the author Biswas (2004), who showed some 35 different ways to look at IWRM. This broad and abstract language creates uncertainty in its interpretation. The word *integrated* communicates the need to consider complexity in policy, so its origins go to concepts such as *integrated planning, integrated environmental management, integrated risk management,* or integration in other policy sectors.

There is no real consensus about the definition of IWRM. My definition is: "Integrated water resources management is a framework for planning, organizing, and operating water systems to unify and balance the relevant views and goals of stakeholders."

This definition is similar to the definition of TWM but is not as explicit on the concept of stewardship. In my opinion, TWM has the advantage in wording because while *integrated* emphasizes blending together, *total* sweeps in the concept of *comprehensive* as well as *integrated*. This may seem like splitting hairs, but unless we use precise terms, each group goes back to the drawing board to create another definition.

Some may confuse IWRM with related concepts, such as Integrated Resource Planning (IRP), a concept developed within the electric power industry with the Energy Policy Act of 1992. It requires utilities to compare all practicable supply options, present an action plan covering a minimum period, designate least-cost options, describe efforts to minimize adverse environmental effects, provide for full public participation, include demand forecasts, and validate predicted performance to assess results (Western Area Power Administration, 2006). IRP focuses on new supply development, whereas TWM and IWRM are broader and deal with stewardship and management in all decision processes.

In 1993, the World Bank (1993) issued a policy paper about a "comprehensive policy framework" that advocated elements of IWRM. The

Bank also issued a "strategy" for water resources management, with an institutional framework (legal, regulatory, and organizational roles), management instruments (regulatory and financial), and a focus on infrastructure and watershed protection (World Bank, 2004).

The Bank's policy paper reflected global attitudes emerging from the 1992 Earth Summit in Rio de Janeiro, and it identified three sustainability principles for water resources management: the ecological principle (requiring integrated management by river basins with a focus on land and water with environmental needs); the institutional principle (all stakeholders participate, including government, the private sector, and civil society); an emphasis on the participation of women; a focus on subsidiarity (actions taken at the lowest appropriate level); and the instrument principle (water is a scarce resource, necessitating the use of incentives and economic principles to improve allocation and enhance the total value from water use).

> **Integrated Resource Planning focuses on supply development**

TWM is consistent with the Bank's emphasis on a comprehensive framework, cooperation and coordination, a reliance on market mechanisms, decentralization and grassroots participation, capacity-building, full-cost pricing with social rates where needed, public or private enterprise, a watershed focus for planning and management, partnering to achieve role coordination and regional cooperation, and a dedication to stewardship of the environment.

Another concept was holistic water management (Kirpich, 1993), an approach that focused on water management in developing countries. It focused on the irrigation sector with an emphasis on interagency coordination, performance standards for water users and staff, the use of indigenous knowledge, local participation for corollary activities, top-down and bottom-up coordination, and the linkage between water and agriculture policy.

Given the incentives to invent new phrases, it seems certain that we will continue to introduce new language for the concepts within TWM. Meanwhile, putting it to work requires sweeping past the rhetoric and getting on with the implementation.

European Union Water Framework Directive

The European Union's (EU) Water Framework Directive (WFD) contains many of the thoughts behind TWM (European Union, 2007). The WFD is a legal framework to assemble several water regulatory programs, with a focus on river basin planning. According to the EU, the WFD establishes "a community framework for water protection and management," which provides for "the identification of European waters and their characteristics, on the basis of individual river basin districts, and the adoption of management plans and programmes of measures appropriate for each body of water."

The EU's goal is to provide for the "management of inland surface waters, groundwater, transitional waters [estuaries], and coastal waters in order to prevent and reduce pollution, promote sustainable water use, protect the aquatic environment, improve the status of aquatic ecosystems and mitigate the effects of floods and droughts."

The WFD has provisions similar those in force in the United States but is seeking a higher level of command and control than is practiced here, which leaves many of these decisions to the states. *Command-and-control* is the opposite of the *laissez faire* approach in the sense that the government or someone tells you what to do and controls your actions, and you are compelled to obey. Table 3-2 compares how features of the WFD are handled by the EU and in the United States. Bear in mind that the jury is out on whether the WFD will succeed completely, as many of these requirements have deadlines well into the future.

An expanded TWM definition

The AwwaRF group introduced and debated many ideas about TWM, and the definition that emerged from their work achieved consensus. Thus, adding concepts will probably not improve on it much. Nevertheless, the group that created the definition focused on issues their organizations faced and the tasks at hand; it did not include concepts of regulatory control or of the responsibility of civil society as stewards of water. Another issue is the responsibility of water service organizations to provide water services to their customers at the same time that they practice TWM.

To recognize these needs and to be explicit about roles and responsibilities, as well as to ensure that TWM includes stewardship of related land resources, I propose two corollaries to the basic definition:

By giving first priority to providing services to their customers, water service organizations build credibility and support for Total Water Management.

Total Water Management requires stewardship by all sectors of society and involves water and related land resources. It requires each organization and citizen to comply and exceed compliance with laws and regulations covering practices that enhance water quality, quantity, and ecosystems.

CHAPTER 3 VISION, PRINCIPLES AND EXAMPLES

Table 3-2. EU Water Framework Directive compared to the US framework

Element of the EU's WFD	The United States approach
Identify all river basins and assign them to national or international districts. Designate a competent authority for river basin districts.	US Water Resources Planning Act (WRPA) specified this in the 1960s but was abolished in 1981. Now on an ad hoc basis.
Complete an analysis of each river basin district, a review of the impact of human activity on the water, an economic analysis of water use, and a register of areas requiring special protection.	On an ad hoc basis.
Identify all water bodies used for human consumption providing more than 10 m^3 a day or serving more than 50 persons.	Source water assessment is part of the US Safe Drinking Water Act.
Produce management plan and program for each river basin district.	On ad hoc basis.
Prevent deterioration, enhance and restore bodies of surface water, achieve good chemical and ecological status of such water, and reduce pollution from discharges and emissions of hazardous substances.	US Clean Water Act (CWA) addresses this.
Protect, enhance, and restore all bodies of groundwater, prevent the pollution and deterioration of groundwater, and ensure a balance between abstraction and recharge of groundwater.	US does not have an overall groundwater policy, but its protection is required under various laws and regulations.
Preserve protected areas.	US does not have this requirement.
Encourage involvement of interested parties in implementation of directive, in particular river basin management plans.	Public participation is required by a number of US laws.
Ensure that water pricing policies provide adequate incentives to use water efficiently and that sectors contribute to recovery of costs of water services including those relating to the environment and resources.	US does not have formal requirements for water pricing.
Publish reports on implementation of the WFD.	US has requirements for reporting under separate acts, such as the CWA.

These additions include the following key concepts:
- First priority to the basic mission of the water service organization
- Stewardship by all sectors of society
- Integration of management of water and land
- Recognition of the role of laws and regulations in TWM
- Going beyond compliance with laws and regulations
- Promoting practices that enhance ecosystems as well as water

What the definition tells us about TWM

The AwwaRF definition (page 58) gives us insight into the elements of TWM and provides answers to the questions of what is TWM, why it is needed, and how it is done.

- TWM is needed because the supply is renewable, but limited, and should be managed on a sustainable use basis.
- TWM is stewardship for the greatest good of society and the environment and balances competing uses through efficient allocation, promotion of water conservation, reuse, source protection, and supply development.
- It enhances water quality and quantity; addresses social values, cost-effectiveness, and environmental benefits and costs; and fosters public health, safety, and community goodwill.
- TWM uses dynamic planning and management that adapt to changing conditions and local and regional variations, with participatory decision-making on a natural water systems basis through a process of coordination and conflict resolution.

Beyond the definition: putting TWM to work

If we base the principles and practices on the definition of TWM, we maintain integrity in the concept. Otherwise, when principles are pulled from thin air, they lack foundation and are continually reinvented in different presentations, much like business books that repackage old concepts with new jargon.

Principles of TWM

The TWM definition gives life to its principles and practices. Table 3-3 takes the basic elements from the TWM definition and converts them into an outline of points for further discussion. Figure 3-4 illustrates TWM in terms of distinctions between what is process and what is principle.

Table 3-3. TWM processes, principles, and practices

TWM element	Process/principle/practice
• Manage on the basis of natural water systems and sustainable use • Encourage planning and management on a dynamic basis that adapts to changing conditions and local and regional variations	• Set effective policies • Plan for sustainable development on a watershed basis • Develop an effective TWM process for planning, decision-making, monitoring, and adapting to change

Table 3-3. TWM processes, principles, and practices, *continued*

TWM element	Process/principle/practice
• Require participation of all units of government and stakeholders • Conduct decision-making through process of coordination and conflict resolution	• Organize shared governance • Define roles and relationships • Commit to a coordination mechanism and rules for consensus and conflict resolution • Implement transparency and accountability
• Balance competing uses through efficient allocation	• Implement system to allocate water resources efficiently and equitably among competing uses • Use incentives for conservation and Best Management Practices (BMPs)
• Address social values, cost-effectiveness, environmental benefits and costs	• Set shared economic, environmental, and social goals • Use effective assessment and Triple Bottom Line (TBL) reporting for unit, regional, and public responsibilities • Reach out with the Corporate Social Responsibility (CSR) program
• Additional elements to expand definition of TWM	• Serve customers first • Regulate effectively • Enable and encourage workforce and public

Process	Set effective policies	Organize shared effort	Set shared goals	Develop a TWM process	Allocate water uses	Regulate effectively	Enable and encourage	Report to the public
Principle		Commit to shared governance.	Set economic, environmental, and social goals.	Plan on watershed basis.	Use fair allocation mechanism.		Use incentives for conservation.	Use TBL reporting.
		Define roles and relationships.		Manage for sustainability.			Educate public and water workforce.	Use effective assessment tools.
				Develop coordination mechanism.				Commit to transparency and accountability.
				Have rules for consensus and conflict resolution.				

Figure 3-4. TWM processes and principles

Set effective policies

Ultimately, what is accomplished by efforts of government or business depends on the policies of each. A *policy* is a plan or course of action intended to accomplish some result. For example, the Clean Water Act establishes national policy goals in the water pollution control sector. The set of policies required to support TWM covers each sector of water management, each government level, and each function of planning and management, so there are many policy categories to consider. Chapter 4 provides more detail on them.

Success in TWM does not depend on government action alone, but having the right policies is a starting point. Policy is important, but it must be backed up by commitments. Usually, these mean funding for government programs, such as enforcement of an environmental law, for example. Some commitments to policy must be voluntary, however, and for these to occur, adequate incentives must be in place.

The policy process occurs in both the legislative and executive branches of all three levels of government, and policy unfolds through laws, regulations, executive orders, and appropriation of funds for program support. Policy is so important that each sector of the water industry has organized efforts to influence it through government affairs committees and similar groups.

Water quality offers an ongoing example of policy development in the United States and is a good example of how difficult it is to create strategies that work in a complex, real-world environment. As water quality policy evolved in the 1960s, advocates of both regulatory-based and market-based approaches made their cases. In the regulatory approach, regulations are set and dischargers must meet them on the basis of command and control, whereas a market approach uses charges and incentives to discourage pollution and to collect revenues to invest in cleanwater facilities. The "right to pollute," limited by a stream's assimilative capacity, could in theory be traded among bidders.

While the regulatory approach seems to go against the grain of free enterprise, it was adopted and has evolved over a number of years. The reasons for the United States selecting the regulatory-based approach were based on water quality objectives, the goal of state primacy and decentralized administration, and a goal to equalize water quality standards across the country (Grigg and Fleming, 1980).

Compared to the regulatory approach, the market approach offers economic efficiency, innovation, and administrative simplicity. However, the regulatory system seemed more practical to policy makers, and it involved no new taxes on industry. The market approach is used in some countries, notably Colombia, and studies show that it does not live up to its promise. In that country, wastewater treatment is inadequate and much controversy surrounds the setting of effluent charges for cities and industries (Grigg et al., 2004).

Limitations of both approaches are evident. As policy evolves, we will probably see combinations of command and control, self-regulation, and the use of economic instruments such as pollution taxes, tradeable permits, and the removal of distorting subsidies. Meanwhile, incentives for voluntary stewardship remain very important.

Organize shared governance

As explained in chapter 1, unless there is a shared effort, the "it's not my problem" syndrome will defeat the quest for sustainability. The mechanism to forge such cooperative approaches is called *shared governance*.

Shared governance is explained in some detail in chapter 4. Basically, it is agreement to share authority so that a decision that is good for all can be reached. Shared governance is essential to avoid the high costs and unsatisfactory results associated with confrontational decision-making. In TWM it is a mechanism to involve all stakeholders in a balanced outcome.

A commitment to shared governance often may be difficult to achieve because of mistrust of government or neighboring communities or both. To illustrate the difficulty, think about the problems we have in achieving regional cooperation and intergovernmental coordination in all types of public-sector problems.

While these barriers are formidable, the path to TWM lies in perseverance in seeking shared governance, and this requires both statesmanship by leaders across the board and a commitment to developing win-win solutions. Many times, the success of shared governance depends as much on good personal relationships among elected and appointed leaders as it does on formal instruments such as intergovernmental agreements and contracts.

Good policy with government commitment is a starting point for TWM

Tampa Bay Water illustrates an instance of shared governance over a regional water supply. There, the Florida cities of St. Petersburg and Tampa, along with the counties of Hillsborough, Pinellas, and Pasco, created a new raw water authority in 1974 through an interlocal agreement under legislation that encourages regional water supply development. Adams (1998) explained how the Florida Legislature found that "cooperative efforts between municipalities, counties, water management districts and the department of natural resources are mandatory in order to meet the water needs of rapidly urbanizing areas."

Adams also explained how the relationships between the agency and its member governments is based on both cooperation and contracts. Regulatory authority is in the Southwest Florida Water Management District, one of the five water management districts in Florida. Tampa Bay Water and member governments must obtain water use permits from the District.

Adams describes how when the authority was formed in the early 1970s, the main problem was development in areas without freshwater supplies. The authority developed additional supplies but has faced a

68 TOTAL WATER MANAGEMENT

number of conflicts in recent years. These led to actions to identify ways to work out conflicts, including a research project to prepare a decision process with trade-off analysis (Tampa Bay Water and CH2M HILL, 2006).

Shared governance is required among the parties in Tampa Bay Water's system, and its board is made up of representatives from member governments. The utility is also quite sensitive to environmental needs and state and federal regulations. In its research project, Tampa Bay developed a multiattribute analysis (MUA) tool called the Source Management and Rotation Technology Tool (SMARTT). The tool enables the utility to work with stakeholders and apply its MUA approach to find optimum solutions to the selection of water supply sources.

Define roles and relationships

Defining roles and relationships means to specify the functions and responsibilities of the parties in a joint venture. Given the many participants in TWM, different groups can be involved in different ventures. Like other businesses, water service providers have missions to serve their customers and to fulfill their corporate social responsibilities (Figure 3-5). A number of government agencies support or regulate water service providers. The support sector provides innovation, services, and products to help the functions of service delivery and regulation. The public also has

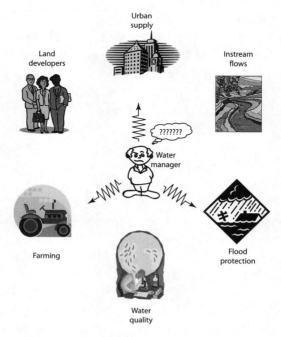

Figure 3-5. Water manager's dilemma

a number of roles, especially to participate in decisions and to be good stewards themselves.

In shared governance, clear delineation of the roles and relationships of the parties is essential, because without it, the only recourse the parties have is to fall back on the "it's not my problem" syndrome.

Defining roles and relationships is in some ways like preparing a contract. The more detail with which they can be defined, the better. That way, there is less room for ambiguity and misunderstandings. In most cases, however, the roles and relationships are not written down so much as understood, having developed historically. This places them in the category of an *institutional arrangement*, meaning they involve historic relationships and are so part of the culture of water management.

In a formal sense, responsibility for defining roles and relationships lies with the convening authority. This might for example be a group of water managers who meet and agree to work together from mutual interests or it might be a state or federal government. When there is no convening authority, say in a case in which irrigators get together on an informal basis to cooperate, roles and relationships may simply evolve.

For example, a mutual aid agreement among agencies might define roles and relationships for operations during disasters. An example of an agreement was prepared for the Florida Panhandle area in 2004 after Hurricane Ivan created serious problems for utilities and the county health department. During the hurricane, communications breakdowns rendered the county's emergency operations center unable to determine the integrity of water systems, forcing the health department to issue precautionary boil water notices. The Mutual Aid Agreement was created to organize the sharing of resources and personnel to help the parties manage water and sewer utilities during and after a disaster (Sims and Kotchian, 2006).

Serve customers first

Unless a water service organization serves its customers well, it will not protect public health and manage water well. So even though customer service does not seem to be an environmental goal, it really is, and it is certainly a social goal. By paying attention to business, a utility builds a base from which it can practice TWM.

The importance of customer service is recognized throughout the water industry and does not need to be explained here. The purpose of including it as a principle of TWM is to emphasize that, without it, a water service organization will not gain support to practice TWM.

Set shared goals

The definition of TWM outlines general economic, social, and environmental goals. These include: to manage on a sustainable use basis; to address social values, cost-effectiveness, and environmental benefits and costs; to enhance water quality and quantity; and to foster public health, safety, and community goodwill.

Arriving at shared goals is essential if people are to work together toward the same ends. For this reason, achieving *consensus* for regional water management actions has become a central goal of shared planning efforts.

The goals should be expressed for every organization, project, and program so that performance measures can be identified for monitoring and assessment. The goals must be shared across an organization and its stakeholders, as well as among different organizations in a region relating to water management. The goals for economic, environmental, and social targets must be set by regions and communities so that they can be translated into specific actions and measures. This is the purpose of Triple Bottom Line (TBL) reporting.

Each entity must set its own goals according to its business purpose and its other responsibilities. Setting joint goals is a shared responsibility of utilities and their partners. Creating a coordination mechanism to establish the shared goals is a regional responsibility that will involve a number of stakeholders and agencies.

For example, threats to estuaries have spurred a number of actions intended to reduce their deterioration. In its legislation for the National Estuary Program, Congress observed that "a proper response will take a cooperative, national effort." It has proven true that coordinated efforts, with each party taking a role, are necessary for estuary programs.

> Coordination means to bring harmony among the stakeholders

The US Environmental Protection Agency (USEPA 1988) prepared a primer to explain the cooperative work needed, and each program was charged to develop a Comprehensive Conservation and Management Plan. The primer summarized lessons learned from USEPA's Chesapeake Bay and Great Lakes programs and the required "collaborative, problem-solving approaches to balance conflicting uses," combining the planning process with politically realistic approaches. It is summarized briefly by USEPA: "the program is woven together by two themes: progressive phases for identifying and solving problems and collaborative decision making."

The process involved the governor's nomination of an estuary program; estuary characterization; the convening of a Management Conference by USEPA, which would produce the Comprehensive Conservation and Management Plan; implementation plans; and continuous monitoring.

The Management Conference would be the place to set shared goals. It is a "forum for open discussion, cooperation, and compromise that results in consensus." The committee structure of the conference targets four groups: elected and appointed officials from all governmental levels, agency environmental managers, local scientific and academic communities, and private stakeholders.

Wisdom from the estuary program has been merged into the *Community-Based Watershed Management Handbook* prepared by USEPA (2005). It describes four steps that lead to a cooperative approach: establishing governance structures according to watershed boundaries, using science to develop and implement a management plan, fostering collaborative problem-solving, and informing and involving stakeholders to sustain commitment.

Develop an effective TWM process

This principle responds to the part of the TWM definition that calls for "planning and management on a dynamic basis that adapts to changing conditions and local and regional variations." In other words, the planning and management processes evolve to adapt to changes that occur.

It is obvious that, for TWM to succeed, an effective process for planning, decision-making, monitoring, and adapting to change is required. In the absence of dynamic actions that occur on a coordinated basis, the regular governmental and legal processes take over, and the "it's not my problem" syndrome will normally govern. The TWM process can be variable to adapt to a particular situation, whether it involves regional water supply, a water quality issue, management of groundwater, or some other water problem.

To create a TWM process, someone needs to take action. The process should be initiated when there is a proposed use or action. Stakeholders are notified and given the opportunity to make input. An authority will be designated to decide, and a legal system for appeals will be in place. All of this is familiar from the planning-and-management process, but TWM demands more effort to involve stakeholders and consider all views.

Elements that make this process work well include shared-vision planning, identifying problems correctly, laying out good alternatives, using fishbowl planning, involving stakeholders, weighing choices, and doing sensitivity analysis. The process must be based on transparency, respect for property rights and the public interest, checks and balances, and other essential principles.

Often no one will have direct responsibility to establish a TWM process, and a water authority or another stakeholder may have to take the lead to get one established. It is difficult for one stakeholder, such as a water authority, to take the lead to establish the process, because others may not follow. It is better if a neutral party not at interest takes the lead, but then you have the problem of creating an incentive for that party to get involved. Nevertheless, the role of leading the process is essential, and someone must take it on.

An environmental impact statement is a way to create a TWM process, and this was the intent of the National Environmental Policy Act (NEPA). NEPA was passed in 1970, establishing goals for environmental policy and requirements for environmental impact statements (EIS) for major federal actions that affect the environment. An EIS evaluates the environmental impacts of a proposed action, unavoidable adverse environmental effects, and alternatives available to the proposed action. In preparing an EIS, the agency consults with other federal agencies with expertise on environmental impact. Since NEPA was passed, the EIS process has influenced many projects and actions. On the positive side, it provides for coordination of diverse interests and improves planning. On the negative side, the process can be bureaucratic, expensive, and time-consuming. TWM is going to be more expensive than the alternative, which is planning that is less comprehensive and that does not look as hard for sustainable alternatives.

The Two Forks Project in Colorado is a case study of use of an EIS in a conflict between water utilities and environmentalists. Two Forks was a large water supply project that had been planned by Denver and was vetoed by the Environmental Protection Agency in 1990. Several aspects of the Two Forks case are of interest. One topic, planning for water supply along the Front Range of the Rockies, will be an issue for many years as a result of rapid urbanization, scarce water supplies, and conflict over water. Another topic is regional cooperation to hold water purveyors together. Other topics included: financing, the roles of players, lack of consensus among the public, the high cost of the EIS, national political agendas, implications of using Section 404 of the Clean Water Act to veto water projects, and the project's aftermath.

Ideally, a TWM process should be collaborative. The Two Forks process was supposed to be that way, but opponents bolted from the mediation process to seek a veto under the EIS process. Under the US legal system, the EIS is the forum for working out disputes between water developers and environmentalists, and in this case, it resulted in a high-profile veto.

Plan for sustainable development on watershed basis

The watershed principle has several parts: planning for sustainable water use, planning on a watershed basis, and using a watershed plan to coordinate water actions in the basin.

Managing water for sustainable development seems like a value rather than a principle, but it can also be a benchmark and a test of plans. If it is used as a test, then the criteria that a plan passes are that ecosystems and water quality are not harmed by the water management action. This principle includes preventative measures such as pollution prevention, minimization of diversions, and environmental management of water storage.

This principle is important because it sets the benchmark and the test for plans. People say they want sustainable development, but by setting this as a principle, we force ourselves to develop measures for how well we do in achieving it. So the importance of this principle lies in the fact that it forces measures whereby water management plans and actions can be tested against sustainability.

Someone must lead in developing a TWM process

As a general guideline, planning for sustainable water use is embedded in the TWM definition. It means to plan to meet the needs of natural systems and not stress them beyond their natural capacities. As a simple example, a project to divert water from a mountain stream will damage natural systems if it leaves too little water to sustain the natural fishery. If the principle is to serve as a benchmark and require measures, then it requires indicators of sustainability.

Water conservation and the use of Best Management Practices (BMPs) are tools to promote sustainable use. Source protection and supply development should also enhance water quality and quantity to protect natural systems. Nonstructural approaches, minimizing diversions, and preventing pollution are all general strategies that promote sustainable use and all are part of the TWM definition.

Planning on a watershed basis is also part of the TWM definition and is more specific than the general goal of planning for sustainability. Planning on a watershed basis requires an arrangement to pull the stakeholders together and can be difficult politically.

The coordination that occurs from having a watershed plan is a result of planning, but it does not occur directly unless someone makes it happen. Here is where the adaptability feature of TWM comes into play. TWM should consider "local and regional variations" as it "encourages planning and management . . . through a dynamic process that adapts to changing conditions." This is a good principle, but it can fail unless you have the right mechanism in place to make it happen.

Unless you plan and manage on the basis of the watershed, how would you ever allocate water and waste loading, apportion benefits from projects, or work together on a watershed sustainability program? It is certainly logical to do this, but to plan on the basis of the watershed is difficult. Two famous men, Jacques Cousteau and Abel Wolman, told us about the difficulties. Ecologist and explorer Cousteau warned that there are enormous political disincentives to collective action and interjurisdictional cooperation (Chinchill, 1988). Wolman (1980), a notable twentieth-century environmental engineer, said, "basin approaches come into criticism by some on the score that basins are essentially non-economic or social units. Viewed by themselves, they represent artificial spheres of action irrelevant to society's needs. The engineer-planner finds them convenient, because he sees them as continuous hydrologic worlds."

The principle of planning and managing for sustainability on a watershed basis is easy to conceptualize but difficult to do. Probably the most difficult obstacle is to get the political actors in a watershed to work together. This is true from the smallest watersheds to the largest river basins.

Recognizing the need for watershed planning, the United States started government programs early in the twentieth century, and by the 1930s New Deal era, the nation was preparing plans for a number of large basins. New Deal thinking about watershed planning found its way into the Water Resources Planning Act (WRPA) of 1965. This well-intentioned program was terminated in 1981, and today in the United States a variety of institutional arrangements exist for planning and management within watersheds. These range from large and formal organizations to small ad hoc watershed organizations. Rather than a centralized and government-mandated approach like the WRPA prescribed, a patchwork of ad hoc arrangements exists. In fact, the United States has several thousand grassroots river and watershed conservation groups which, along with local agencies and government, provide a great deal of planning and coordination (River Network, 2006). This does not imply that the existing arrangements are adequate, because many water projects and actions do not enjoy the coordination envisioned by TWM principles.

TWM leads to sustainable watershed management

The question of who has responsibility for watershed planning and coordination is a vexing one because of the points made by Wolman (1980) that basins are noneconomic or social units, are artificial spheres of action irrelevant to society's needs, and are convenient mainly to engineers and planners. In other words, the way the politician sees things is different from the way the planner sees them.

Planners lack the authority to bring people together. Politicians have the authority but in some cases act only when it is urgent or when self-interest is involved. When politicians focus on the public interest and go beyond political expediency, they may get worn out with the long durations and hard knocks of water negotiations. So the institutional issues that deter implementation of the basin planning principle are formidable.

These institutional obstacles can be overcome by developing ad hoc and working arrangements at the water manager level. This also requires a focus on the public interest and going beyond the "it's not my problem" syndrome.

Watershed-based planning is easy to conceive but difficult to do. Candidate watersheds can vary in size from a few square miles to large basins covering whole regions of the country. In the smaller basins, the issues might be limited to stormwater management and nutrients that impact a small lake. The larger basins can involve water issues and stakes with billions of dollars of impacts on regional economies.

On the small watershed end, a local planning group for water quality in the Big Thompson River watershed in Colorado has established a monitoring and assessment program for water quality. The group, called the Big Thompson Water Forum (BTWF), is one of the thousands of watershed groups around the country. It has a board of directors, a group of participating utilities, and membership in the basin. It is the main planning group to coordinate matters within the some thousand square miles of the basin.

On the other end of the spectrum of sizes, the Great Lakes form a mega-basin that requires planning on a scale thousands of times as large as the BTWF. Institutional agreements on the Great Lakes trace back to 1909, when the United States and Canada signed a Boundary Waters Treaty and established the International Joint Commission (IJC). Over the years, additional agreements have been signed, and today the numbers of players in this effort are multitudinous. Coordination is a tremendous problem and even has international challenges. In 2005, a strategy was developed by the Great Lakes Regional Collaboration of National Significance, a partnership of key stakeholders, to work together toward a goal of restoring and protecting the Great Lakes ecosystem (USEPA, 2007).

Commit to coordination

A coordination mechanism brings interests together to achieve a harmonious result. It is in ways like lubrication between hard surfaces. With it, you get a smooth result. Without it, there is friction and a wearing down of surfaces. In TWM, we need to balance among competing interests, and coordination is the way to do it. However, coordination does not just happen by itself; it has to be energized by some mechanism.

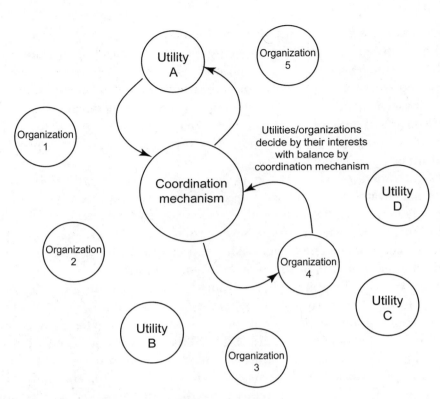

Figure 3-6. Coordination and balancing in TWM

Examples of such mechanisms include committees, facilitated meetings, a web page to disseminate information, or an individual to communicate among the stakeholders. Figure 3-6 illustrates the kind of coordination required by TWM.

If there is no coordination mechanism, then everyone goes his own way and there is no possibility for integration. Water management, by definition, requires coordinated approaches to provide the balance needed. The need for this is evident in the TWM element of balancing competing uses of water.

Actually, coordination is a complex phenomenon involving a number of requirements. The need to have rules for consensus and conflict resolution goes with the coordination mechanism: if it is to work, the rules are required. Management itself has a coordination function built into it. If management functions are to plan, organize, and control, you might think of coordination within these as getting everyone on the same page and conducting the orchestra.

Coordination is everyone's responsibility, but you need a designated and formal way to bring it about. Think of it like marriage counseling. Two people who need to work a few things out each have responsibility for his or her part of the coordination. They might need a formal mechanism in the form of counseling to bring out the issues, arrange compromises, and smooth over the rough points. In the same way, the partners in a water management venture should work together to work out the coordination mechanism that works for them.

When water flows across a state line or an international border, it creates a transboundary issue. Examples of this situation in the United States are the Colorado River, involving seven states and Mexico; the Delaware River, involving New York, Pennsylvania, Delaware, and New Jersey; and smaller rivers such as the Platte, involving Colorado, Wyoming, and Nebraska.

Within unified political jurisdictions, these can be handled by river basin plans, but when more than one state is involved, the complexity grows. Thus a coordination mechanism as an institutional arrangement is required to coordinate among agencies and interest groups in the basin. The mechanisms vary with the institutional settings and the physical and economic conditions in the basins. In simple cases, these might range from informal networks among water managers to formal commissions, authorities, and water management districts. Most common are ad hoc attempts to coordinate. For example, in Colorado a South Platte Coalition of agencies and interest groups was formed. In the 1980s, Colorado's governor formed a Front Range Water Forum to cover water supply issues.

Coordination mechanisms for interstate streams involve different forms. Kenney and Lord (1994) summarized the historical development of these mechanisms and identified seven different categories: the interstate compact commission, federal-interstate compact commissions, the interstate council, the basin interagency committee, interagency-interstate commission, federal regional agencies, and a single federal administrator.

Allocate water resources efficiently and equitably

The goal of this principle is to balance "competing uses of water through efficient allocation that addresses social values, cost effectiveness, and environmental benefits."

What are competing uses? Chapter 2 explains that water is used for supply for cities, industries, and farms; to carry wastewater; for instream purposes, including hydropower generation, recreation, and navigation; and to nourish the natural environment. These are the main uses that compete with each other. Note that it is the *resources* that are allocated, not only the direct use of water for a supply. In other words, if water is used for purposes such as hydropower, environmental enhancement, or

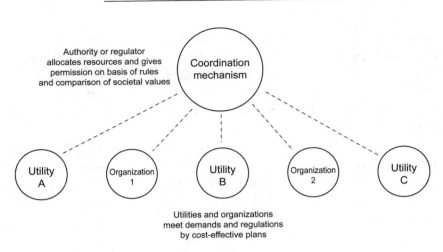

Figure 3-7. Coordination and allocation of water

waste assimilation, the resources of the water are still allocated, even if the water is not actually consumed.

The allocation occurs at different levels, as explained before. Figure 3-7 illustrates how after an authority allocates the overall resource, then each organization makes its own allocation to achieve its purposes.

This principle also addresses the TWM element of addressing social values, cost-effectiveness, and environmental benefits. The dilemma is: who decides on the balance among them?

Lack of balance and equity hurts all interests. To see that issue, consider the many examples of unsustainable development that have occurred. A quote for that is contained in the report of the Longs Peak Working Group (1992), a committee that recommended water policy changes after the 1992 election. They wrote, "Sound water policy must address the contemporary and long-term needs of humans as well as the ecological community. Nationally, we have not been using water in a manner to meet these needs on a sustainable basis. Examples include the endangered Columbia River salmon, the overtaxed San Francisco Bay Delta, the poisoned Kesterson National Wildlife Refuge, the salt-choked Colorado River, the vanishing Ogalalla Aquifer, Louisiana's eroding Delta, New York's precarious Delaware River water supply, and the dying Florida Everglades." As you can see, the group chose adjectives to dramatize the problems, but more than 15 years after that report, we can see that none of these problems have been resolved and some are worse.

While allocation of water rights and permits is part of this principle, it is not all of it. Allocating the resource involves much more, including items like the assimilative capacity of water to handle pollutants. The goal is to have equitable decisions to balance all competing uses of water through efficient allocation that maximizes the TBL. The ideals portrayed in this goal are worthy, but the challenge is in implementing them.

For water supply allocation, states either work on permits or on the appropriation doctrine, the latter which is used in Western states. To allocate water quality assimilation capacity, the Clean Water Act and all of its rules are applied. To manage instream flows for supply, hydropower generation, recreation, and navigation, a group of laws apply, such as the Federal Power Act. To assess environmental needs and the requirement for instream flows, a patchwork of state laws applies, but these are of limited effectiveness.

Consensus should be used wherever possible, but when it is not possible, a coordination mechanism might resolve conflicts, or ultimately the regulatory or legal system might have to decide. Conflicts should be managed with negotiation and alternative dispute resolution, if possible.

To illustrate the range of issues involved here, consider the ideas contained in a *Wall Street Journal* (1991) editorial about water problems in the Northwest. The editorial is rich with ideas but short on practical details, and useful for class discussions about water management. The editorial discusses how to improve conditions for the salmon and trout that cannot reach their spawning grounds because they are blocked by dams that provide hydroelectricity, irrigation water, and navigation. The editorial opines: "Trying to even nature's score is an understandable aim," and "we're not much closer politically to making the proper trade-offs today than we were when the big dams went up early in this century." Making the proper trade-offs means allocating the resource. The dilemma, according to the editorial, is that resource trade-offs made by judges may be based on the findings of some "apocalyptic scientist" who does not have to consider jobs.

> **Allocating water resources addresses all uses of water**

The solution, according to the editorial, could be a free-market exchanging of water rights instead of dividing them up politically. Those who would presumably purchase the environmental rights would be groups such as Salmon Unlimited and state legislatures. They would be willing to let a species disappear if no one would pay for its survival. Their concluding statement summarizes well: "Approximating a free market in natural resources isn't going to be easy—especially when so many parties have careers and causes at stake. But it's hard to think of any other mechanism capable of arbitrating the myriad demands of millions of people in an economy."

Although this editorial comes from a conservative paper and looks more at the economic side than the social and environmental sides of the case, it explains the dilemma well.

Regulate effectively

Regulation means controlling behavior in accordance with a rule or law and is aimed at controlling activities to protect the public interest when private markets do not. While voluntary coordination is a key element of TWM, it is a fact that regulation is needed to force us to take some actions that we would not otherwise take.

This principle in the water industry deals with health and safety, water quality, fish and wildlife, quantity allocation, finance, and service quality. These are the same economic, environmental, and social arenas that require balance in TWM. So to regulate effectively means to use regulation when other TWM tools are inadequate.

Regulation is necessary in TWM because of the shared nature of the water resource. One person's waste affects another person's drinking water. As water utilities are monopolies, it is impossible to have an unregulated water industry. A saying that captures this is "don't have the fox guarding the chicken coop." It is a version of the separation-of-powers principle in government, although the regulators who write the rules also enforce them.

Having an effective regulatory program is required because people do not always follow the rules, whether through carelessness, indifference, or even malevolent intent. The definition of TWM does not explain how water laws are mostly regulatory and how regulation is essential for the coordination and control purposes of water management. For this reason the enhanced definition of TWM that includes compliance with laws and regulations is important.

Regulation works by balancing the public's interest with those of other interest groups. Interest groups push their agendas through regulations and laws. Each sector of the water industry has its own regulatory programs. Examples include the Safe Drinking Water Act, the Clean Water Act, floodplain regulation, and the National Environmental Policy Act. The controls come in the form of rules, permits, monitoring, and enforcement of regulations.

A regulatory program must have an enforcement mechanism to be taken seriously. To understand enforcement, consider the levels of laws and regulations. First there is the law, or the statute, normally either federal or state. Then there are regulations, such as a regulation about stream water quality standards. Then there are various reports and procedures that are administrative in nature and are needed to implement a program. Any of these can be the subject of an enforcement action. Regulators involved in enforcement range from local health department inspectors to officials of the US Department of Justice.

In my experience as an environmental official, I have observed a few basic principles. The first is that enforcement staff, from top to bottom, need to know the rules of the game. Second, they must know that the system is honest and businesslike. Then they must be sure to have reliable monitoring information to base decisions on. Information must be reported up the channels to officials in a manner so as to initiate study and consideration of enforcement actions. Enforcement officials should try all means of obtaining compliance before actually levying penalties. This can be quite time-consuming and emotionally draining because of the realities of government-business relations. The system of enforcement must be efficient. Finally, appeal panels must be wired into the system in a manner so as not only to exercise justice to the persons or organizations being penalized but also to back up the goals of the environmental program. If an appeals judge doesn't see the dangers in someone throwing toxic chemicals in the water because the judge lacks sensitivity for the environment, programs can't be effective.

There is no comprehensive water industry regulatory policy. The total picture is a mélange of federal, state, and local laws and regulations that govern water service providers and individual water users. Because much of the water services are provided by local government, regulation comes from federal laws implemented by state agencies. Other regulation is informal, through the political process. We often hear calls for "regulatory relief" and "regulatory reform" because people and businesses don't like being "regulated." However, regulation is a price we pay to live together in a civilized society. The challenge is to regulate enough but not too much. Regulation seeks to apply law to control behavior in the public interest, but defining the public interest is an elusive goal. Nevertheless, this is the front lines of the quest for sustainable development.

Regulators are part of the TWM team

Anyone working for the US Environmental Protection Agency is a regulator, although some more so than others. By the same token, the state EPAs or water quality agencies are regulators. The Corps of Engineers has a regulatory component by delegation of the authority to implement Section 404 of the Clean Water Act. The US Fish and Wildlife Service and Forest Service have entered the regulatory arenas with the Endangered Species Act and enforcement of federal water-related rules in National Forests. In the West, state engineer offices are regulators in the sense that they control the diversion of water from streams and wells. Eastern states are increasing their activity in this area. State natural resources departments with dam safety missions regulate various aspects of safety. Similar functions have developed in the East. State public service commissions regulate costs of water service for some utilities. These commissions, where they are concerned with water at all, regulate only private water companies.

A large fraction of the world's population lives very near the coastlines, but one of the toughest jobs in water management is managing water quality and ecological issues in estuaries and coastal waters. These water bodies harbor diverse fish and wildlife species and are deeply appreciated for their beauty. On the one hand, their ecologies are fragile and their productivity is vital to fish and wildlife, but on the other hand, they are vulnerable to the large population and industrial regions located near them.

The importance of estuaries was brought home to me in the 1970s when North Carolina faced the problem of restoring its Chowan River. The state developed a Chowan River Restoration Project, and this work has since been merged into its Albemarle-Pamlico estuary program. The project and its subsequent estuary program involved many stakeholders and participants. One stumbling block to a solution to the river's problems was lack of agreement among North Carolina and Virginia, the local governments, and industries about the solutions needed. The North Carolina governor, James Hunt, requested that the US Environmental Protection Agency (USEPA) assign a scientific team to audit the restoration project and recommend solutions so that regulatory actions could be taken.

Regulatory actions taken to restore the river included stream standards, new regulations on nutrients, and enforcement actions against industries and municipal dischargers. Such coastal water problems remain very difficult, but evidence suggests that the Chowan River improved in water quality following these regulatory actions.

Use incentives for conservation and BMPs

This principle introduces tools for good water management into plans. The TWM definition calls for *promotion* of water conservation rather than to require it, and the way to do this is to use incentives. The TWM definition does not call explicitly for best management practices (BMPs), but it is logical that they are included because they are the way to achieve stewardship in use of land insofar as it impacts water resources. For example, a stormwater BMP will reduce pollution of receiving waters, thus improving its prospects for water supply.

BMPs focus on the twin issues of nonpoint sources and hydromodifications, both of which create impacts from land development and uses. TWM encourages use of best management practices to prevent pollution and deteriorated ecosystems. The myriad of small actions from nonpoint sources and hydromodifications have a large cumulative effect on natural water systems and ecosystems, and BMPs can have a large positive influence in mitigating these cumulative effects.

Use of conservation and BMPs is important because they are the main instruments outside of regulation that enable us to protect and exercise stewardship over water resources. Regulation alone is not enough; voluntary protection and stewardship are required to deal with the many small actions that were discussed in chapters 1 and 2.

Conservation and the use of BMPs fall into a general category we might call the *conservation ethic*. That is, if you use anything that you possess carefully and also take care of, then you are conserving it. Conservation and BMPs are applied to our common resources more than to our private resources, so we need incentives to overcome the Tragedy of the Commons.

BMPs can be mandatory in some cases but should be used even when voluntary. The problem when they are voluntary is in getting land developers to spend extra money. Land developers and users, including road departments, have the primary responsibility for implementing this principle. Regulatory control over BMPs and land development is limited, so overcoming institutional obstacles is important to making it work.

Responsibility for BMPs is hard to fix. For example, parking is a real problem on university campuses. My university constructed a three-acre satellite lot to accommodate student vehicles. The capacity of the lot was 500 cars, and the design was simply a field of asphalt marked for the parking spaces. Drainage was to inlets that collected the stormwater and conveyed it directly to a nearby stream that discharged into the local river. Whereas before there were three acres of open field, now the runoff comes faster and is loaded with dust, grease, and the usual load of urban contaminants.

If the university had used BMPs on this parking lot, the cost would have been higher, including a cost for maintenance (otherwise the catch basins, straw bales, or other measures would become unsightly). With the existing design, the lot looks neat, but the pollution problems have been transferred to the river. The incentives, however, pointed toward the plan that was implemented. It would take a culture shift to change a design like this so that the university would not build a facility without BMPs. A TBL report would disclose how the parking lot was adding to river pollution while minimizing student parking cost, but no such report is required. To pick up the monitoring and assessment information for the TBL report, the university's facilities department would have to report on construction, stormwater, and transportation/parking activities.

Many concepts can be introduced for a given region for the use of incentives for conservation and BMPs. A few examples are:
- Utilities use rate structures that reduce in-house use of water.
- Xeriscape landscaping is encouraged in a city.

- Customers are encouraged not to dispose of harmful substances in the water.
- A sustainability report is issued by the utility to emphasize shared responsibility.
- A city's stormwater program emphasizes citizen education for clean streams.
- Farmers receive financial credits for installation of BMPs.

Water meters and rate structures provide an example of incentives for conservation. A case unfolded during the 1980s in Fort Collins, Colorado, when the city council was debating whether to require water meters to replace the old practice of no meters and flat rates based on lot size. Water use at that time was between 200 and 230 gallons per capita per day (gpcd) in an average year. It might seem incredible to some that there was so much opposition to metering, but at that time and place not everyone favored them, by a long shot.

Arguments in favor of metering were that it would induce water conservation and reduce raw water requirements; it would reduce peak day demands and the need for treatment plant capacity; it would allow cost-of-service rates; it would enable demand management, monitoring of system losses, and efficiency in the system; and it would place choice in the hands of the customer. In addition, advocates said, meters would be good stewardship.

Arguments against meters were that they were expensive, it added a cost to read and maintain them, and the quality of green lawns might decline. Another argument, that meters would discriminate against low-income customers, split the council. Some who normally voted for environmental issues voted against meters because of possible discrimination against low-income people. Another argument against meters was that water waste actually constituted a "reserve," and the meter card could be played at any time.

Incentives are needed for conservation and control of NPS

The city water board debated meters often, but the usual result was "no meters, water waste is our water reserve." However, it finally voted to recommend a metering program to council on a split vote. The council deadlocked on the proposal, and it went down to defeat. This did not mean the council was against water conservation, and to prove that, it initiated a "demand management" policy (Clark and Bode, 1993). In 1991, the State of Colorado took matters into their own hands and passed a mandatory water conservation bill that included metering, so Fort Collins had to comply anyway. In 2007, water use in Fort Collins was about 160 gpcd, and the city now uses an inclining block rate to encourage conservation.

Use effective assessment and TBL reporting

The principle of assessment and reporting goes along with multiple goals, and it is the way to counter the Tragedy of the Commons. Assessment means to look at the data and interpret what they mean in the light of your goals. Triple Bottom Line reporting, as explained in chapter 5, means that you report on how you are doing on all of your goals.

Figure 3-8 shows why these are important. As outlined in chapter 1, the cause of the Tragedy of the Commons is self-interest and the cure for it is public interest. However, public-interest actions do not always occur naturally because, although many people are public-spirited, the majority of us are usually pursuing our self-interest. Public-interest actions occur because of the shared values and stewardship inherent in TWM. To stimulate these, we can encourage, cajole, threaten, or take any other actions, but without accountability, no one knows what happens and soon our efforts die away. However, with accountability, we keep our focus and we disclose to others how we are doing. Accountability reinforces shared values and governance.

There is interest among leading utilities in TBL reporting, but it has to be used more effectively and by more entities. Chapter 5 explains in detail how it works and gives examples. Assessment is a separate task and is explained in chapter 7. Basically, the two work together this way. Some entity (a watershed authority, a city or county government, a metro district, etc.) has responsibility for water management in a region. Rather than simply reporting on rates and compliance with requirements under the Safe Drinking Water Act (SDWA), the entity decides to assess the

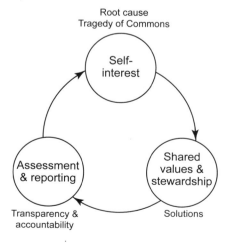

Figure 3-8.   Assessment and reporting to counter Tragedy of the Commons

overall state of water resources in its jurisdiction and to consider economic, social, and environmental goals related to water. It has taken the lead to work with other stakeholders to formulate shared goals, so it has categories in which to assess and report by the TBL.

Without the goals, you cannot assess very well, because you must know what you are assessing. Even a clear-sounding concept like water quality is difficult to assess unless you know its definition, and its definition needs to be related to some goal.

Here is a simple example of assessment and reporting in a local watershed group. The Big Thompson River flows from its headwaters near Longs Peak through several ecological zones to end at the South Platte River near Greeley, Colorado. Since its earlier era as a pristine river and a haven for beaver, the river has seen many changes in its basin. Growth has centered on farming, town development, and the recreation industry along the river's upper reaches, which includes a drainage area in Rocky Mountain National Park. While the basin's population is not huge, the river has been impacted by land uses and in its lower reaches is far from a pristine stream now.

During the 1950s, West Slope water began to be imported through the Adams Tunnel, part of the Bureau of Reclamation's Big Thompson Project. Now, growth on the other side of the Continental Divide affects Big Thompson water, too. Skiers in the Colorado mountains don't realize it, but they are linked through water quality to residents and visitors of the Big Thompson basin.

It's not just a simple matter of water quantity and quality. There is a symbiotic relationship between a river and its riparian corridor. The science of ecology explains how aquatic life and terrestrial life depend on a river's flows. To preserve and manage the Big Thompson requires multiple strategies, and it's easy to see both sides of the question. If nothing is done, the stream will degrade. On the other hand, it is not feasible to return it to completely natural conditions because we rely on it for many purposes, including water storage and diversions. So we must manage the river for multiple purposes, achieving a healthy balance that nourishes environmental conditions while meeting human needs.

How can we assess our success in achieving multiple goals? The Big Thompson Water Forum (BTWF) was organized in the 1990s to monitor and assess water quality, but the goals go beyond simple measures of water quality. The forum conducts a successful monitoring program that is narrowly defined by its municipal supporters, but it faces many controversies in converting the program to a valid overall assessment, much less a Triple Bottom Line report.

Commit to transparency and accountability

Transparency and accountability are important to blunt the negative outcomes of self-interest and to have shared values and stewardship. Transparency means that information is freely available to all stakeholders. This is an important value in the United States and is expressed in many ways, including through our free press and the Freedom of Information Act. Accountability means that if you are responsible, you must present your results for checking by some authority. It's like a report card, with accountability for your child's schoolwork. In the context of TWM, the goal is to provide information on a transparent basis for water leaders and citizens to promote water citizenship, responsible behavior, and stewardship of the environment.

Transparency builds trust

Ideally, assessment and reporting take place at local levels, where you can have grassroots participation to gain support for programs. Regulations for assessment and reporting, such as through the Clean Water Act and the Safe Drinking Water Act, should be viewed as minimum requirements.

As responsibility is hard to fix for assessment and reporting, it follows that transparency and accountability do not happen naturally. Citizen participation and effective governance are the keys to them.

The increasing use of web pages to disseminate water information is a good example of transparency and accountability in water management. Someone has to foot the bill for these; they do not just happen. Once they are created, to sustain them may require a government sponsor. For example, the Chesapeake Bay program has been under way at least since the 1970s, with earlier efforts predating that time. Today, the web page of the program provides considerable information about the Bay (Chesapeake Bay Program, 2007). As it is a partnership program, it has taken on a life of its own. Without government sponsorship and funding, however, it would be hard to sustain its level of activity.

Enable and encourage the workforce and the public

This principle is to equip and encourage both the water workforce and the public to go beyond minimum requirements to exercise across-the-board stewardship in water management. Examples of what this means abound. For example, a water utility manager can exceed rather than just meet standards. A utility crew can avoid dumping sludge into a waterway. Homeowners can avoid disposing of harmful wastes into the sewer system. A homeowner's association can be a good steward of a small creek.

When all is said and done, regulatory controls and incentives only go so far. Some of the responsibility for TWM outcomes still depends on the water workforce and the public doing the right thing when no one is looking. This requires us to be instilled with the right ethics; that is, exercising correct behavior and self-regulation toward water and all of its users and customers. The water workforce is critical to TWM because many actions in water management can be done at a minimum level, or they can be done with care and with good stewardship. The public bears a great responsibility for water stewardship, and that is what the field of environmental education is all about. Stewardship is the shared responsibility of a wide band of players, and no one has sole responsibility for it. At the end of the day, the result of environmental education should be the knowledge and motivation for all to be good stewards.

This is a principle that is easy to state but hard to monitor and enforce. Basically it works by everyone trying a little harder and having integrity to do the right thing when no one is looking. Monitoring and reporting results can help greatly, but in many cases there is no single point of responsibility for corrective action. This aspect of TWM involves many players and small details and is hard to plan. Volunteer actions go a long way in this arena.

It is important to equip the TWM team

Utilities and water management organizations can exercise their external responsibilities by identifying the actions needed and developing ways to measure results so they can be reported to the workforce and the public. Professional associations can implement programs to help. Educators have a big role in enabling the public through environmental education.

Chapter 11 is about environmental education, which is a good example of how the workforce and the public should be enabled to practice TWM and stewardship. Environmental education is any organized school or public education effort to teach about how natural environments function and how to manage behavior and ecosystems to live sustainably. It includes topics across a wide spectrum, and water management involves a subset of them that requires understanding each person's role in sustaining the water environment.

Reach out with CSR program

Corporate social responsibility (CSR), which is discussed in some detail in chapter 11, means for a corporate body (private or public) to reach out to undertake broader social responsibilities that go beyond its direct mission. It does not mean that every business should be a social agency, but it means that a corporate body looks after some aspect of the public interest as well as after its own private interest.

CSR is important because if everyone only looks after his or her private interest, the Tragedy of the Commons operates. CSR is a way to organize the work that an organization can do and to report on it. Not every organization can do everything, but a consulting engineering company can have an outreach program to local schools on water, for example. A utility can sponsor studies on how to improve a watershed, and a local business can send people to be involved in a stream cleanup, for example.

In northern Colorado, the water utilities pour a lot of effort into the Children's Water Festival, an annual event in which elementary school children are bused to a facility where they see a number of fun water displays. This is part environmental education and part science education, and the utilities and their staffs are doing a good bit to motivate and educate the youth about water.

CSR is another name for "going the extra mile"

CSR works when leaders of an organization decide to undertake it. After all, public organizations have legislatively mandated and enabled program responsibilities, and they will undertake them first. Private companies have the profit motive, and they must pursue it or they can't survive. CSR can occur when these basic responsibilities are met, and it will only happen when an organization's leadership decides to implement it.

Summary points

- TWM is a framework of principles, processes, and practices. A principle is a rule for action, a process is a collection of actions that leads to a result, and a practice is a systematic way of doing things, such as a *management practice*.
- TWM and integrated water resources management are similar concepts with the goal of balancing the relevant views and goals of stakeholders. The basic principles of these complex concepts can be presented in practical ways for use by water managers. In this way, they become more than just academic ideas.
- TWM includes the following principles:
    - Set effective policies.
    - Plan for sustainable development on watershed basis.
    - Develop an effective TWM process for planning, decision-making, monitoring, and adapting to change.
    - Organize shared governance.
    - Define roles and relationships.
    - Commit to coordination mechanism and rules for consensus and conflict resolution.

- Implement transparency and accountability.
- Implement system to allocate water resources efficiently and equitably among competing uses.
- Use incentives for conservation and BMPs.
- Set shared economic, environmental, and social goals.
- Use effective assessment and TBL reporting for unit, regional, and public responsibilities.
- Reach out with CSR program.
- Serve customers first.
- Regulate effectively.
- Enable and encourage workforce and public.

Review questions

1. Why is *water management* difficult to define?

2. Explain whether TWM applies to both structural and nonstructural measures and give examples.

3. In your opinion, is the definition of TWM as "stewardship of water resources for the greatest good of society and the environment" a workable concept or simply a visionary phrase?

4. Which principle of TWM do you consider to be most important in achieving sustainable development? Why?

5. Of the principles of TWM, which ones seem most aligned with the political process? What does this mean for water managers?

References

Adams, Alison. 1998. Analysis of Regional Water Conflicts: The Case Study Approach. Ph.D. Diss., Colorado State University, Fort Collins.

American Water Works Association. 1994. Principles of Total Water Management Outlined. *MainStream* (November) 48(9): 4,6.

American Water Works Association Research Foundation. 1996. Total Water Management Workshop Summary. Draft. Seattle, Wash., August 18–20. Denver, Colo.: AwwaRF.

Biswas, Asit K. 2004. Integrated Water Resources Management: A Reassessment. *Water International* 29 no. 2: 248–256.

Chesapeake Bay Program. 2007. http://www.chesapeakebay.net/. Accessed May 26, 2007.

Chinchill, J. 1988. Chesapeake Bay Restoration Program: Is an Integrated Approach Possible? In *Water Policy Issues Related to the Chesapeake Bay*, ed. William R. Walker. Blacksburg, Va.: Virginia Water Resources Center.

Clark, Jim, and Dennis Bode. 1993. *Water Conservation Annual Report*. City of Fort Collins, Colorado. January.

European Union. 2007. *Framework Directive in the Field of Water Policy.* http://europa.eu/scadplus/leg/en/lvb/l28002b.htm. Accessed June 5, 2007.

Global Water Partnership. 2004. *Managing Water.* http://www.gwpforum.org/. Accessed November 20, 2004.

Grigg, N. 1996. *Water Resources Management: Principles, Regulations, and Cases.* New York: McGraw-Hill.

Grigg, N., L. MacDonnell, J. Salas, D. Fontane, L. Roesner, C. Howe, and M. Livingston. 2004. *Integrated Water Management and Law in Colombia: Institutional Aspects of Water Management, Allocation of Uses, Control of Contamination, and Urban Drainage.* Colorado State University.

Grigg, Neil S., and George H. Fleming. 1980. *Water Quality Management in River Basins: US National Experience.* International Association of Water Pollution Research, *Progress in Water Technology* 13. London: Pergamon Press.

Kenney, D., and W. Lord. 1994. *Coordination Mechanisms for the Control of Interstate Water Resources: A Synthesis and Review of the Literature.* Report for the ACF-ACT Comprehensive Study, US Army Corps of Engineers, Mobile District, Alabama, July.

Kirpich, Phil Z. 1993. Holistic Approach to Irrigation Management in Developing Countries. American Society of Civil Engineers, *Jour. Irrigation and Drainage* 119 no. 2: 323–333.

Longs Peak Working Group. 1992. *America's Waters: A New Era of Sustainability.* Natural Resources Law Center, University of Colorado, Boulder, December.

River Network. 2006. *Directory of Watershed Organizations.* http://www.riverwatch.org/. Accessed November 4, 2006.

Sims, D., and S. Kotchian. 2006. *Mutual Aid Agreement for Water and Sewer Utilities* 2005–2006. St. Louis, Mo.: National Environmental Public Health Leadership Institute.

Symons, J.M., L.C. Bradley, and T.C. Cleveland. 2000. *The Drinking Water Dictionary.* Denver, Colo.: American Water Works Association.

Tampa Bay Water and CH2M HILL. 2006. *Decision Process and Trade-Off Analysis Model for Supply Rotation and Planning.* AwwaRF Report. Denver, Colo.: AwwaRF.

US Environmental Protection Agency. 1988. *Estuary Program Primer.* Washington, D.C.

USEPA. 2005. *Community-Based Watershed Management Handbook.* EPA 842-B-05-003.

USEPA. 2007. Great Lakes. http://www.epa.gov/glnpo/collaboration/strategy.html. Accessed May 17, 2007.

Wall Street Journal. 1991. Lox Horizons. May 20.

Western Area Power Administration. 2006. *Integrated Resource Planning Guidelines.* http://www.wapa.gov/powerm/pmirp.htm. Accessed September 28, 2006.

White, G.F. 1998. Reflections on the 50-year international search for integrated water management. *Water Policy* 1: 21–27.

Wikipedia. 2006. *Paradigm.* http://en.wikipedia.org/wiki/Paradigm. Accessed August 26, 2006.

Wolman, A. 1980. *Some Reflections on River Basin Management.* Proceedings, International Association for Water Pollution Research Specialized Conference on New Developments in River Basin Management, Cincinnati, Ohio.

World Bank. 1993. *Water Resources Management: A World Bank Policy Paper.* Washington, D.C.

World Bank. 2004. *Water Resources Sector Strategy: Strategic Directions for World Bank Engagement.* http://www-wds.worldbank.org. Accessed November 12, 2006.

CHAPTER 4

PLANNING AND SHARED GOVERNANCE

Among the principles of TWM, one stands out as extremely important but also extremely difficult. That principle is *shared governance*, or agreements to share authority to reach decisions that are beneficial for all. It is a no-brainer that shared governance is difficult. Among city boards, for example, it is often difficult enough to reach agreement within the group, but much more difficult to reach agreement with another city. Perhaps the most glaring example is the difficulty the nation has with regional government, which distributes services across multijurisdictional areas. A later section of this chapter describes its promises and challenges.

Fortunately, we have a tool to use to work toward shared governance. As with other TWM tools, it sounds familiar but is hard to practice. It is called planning. As the European Union Water Framework Directive shows us, the discipline of *water resources planning* is a centerpiece of TWM. The other topic of this chapter—governance—has been recognized around the world as the key element to make water management effective. So this chapter addresses two central issues of TWM: planning and governance.

Several principles from the definition of TWM deal with one aspect or another of planning or governance. These are:
- Develop an effective TWM process;
- Organize shared governance;
- Plan for sustainable development on a watershed basis;

- Set shared goals; and
- Commit to coordination.

How these principles work through the planning process is shown in Figure 4-1, which illustrates how planning for sustainable development on a natural systems basis is the driver and the TWM process is the response.

Shared governance is essential in TWM, but it is difficult

Without shared efforts through TWM, the "it's not my problem" syndrome may block sustainable water management solutions. But organizing shared efforts is often difficult because of perverse incentives and institutional constraints. Perverse incentives are those that create actions that are the opposite of those that are needed.

This is a problem all through our government system, which has arrangements for shared problem-solving but finds difficulty in making them work. These arrangements include intergovernmental coordination, public-private partnerships, and working through public-interest organizations, among others. In theory, such arrangements can help overcome problems between the public and private sectors, but they are hard to implement.

Shared governance is a mechanism for cooperation and the sharing of power and decision-making. Simply speaking, it is a way to work together to solve common problems. It is a nice theory, but the challenge is how to make it work.

Actually, the mechanism of water resources planning offers a path to cooperation and joint problem-solving through several channels, including shared governance. While the direct meaning of *planning* is to design a project or program, the field of water resources planning takes on a much broader scope. It includes topics from policy development to public participation. Figure 4-2 illustrates how it occurs at different levels.

Shared governance within TWM is the way to balance decisions by people and organizations and to develop win-win solutions. TWM works through water resources planning to organize shared governance arrangements.

The chapter explains how water planning and management should work, with an emphasis on the political dimension of shared governance. For water resources projects, it deals with plans by organized water management units. Where it addresses the myriad smaller impacts from nonpoint sources and hydromodification, it addresses policy planning that leads to rules and programs to encourage stewardship, best management practices, and self-regulation.

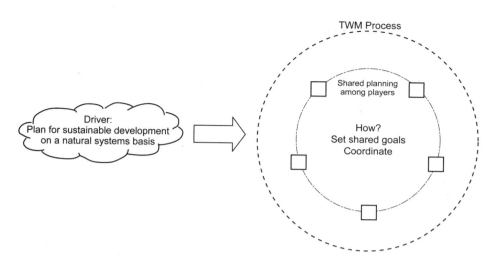

Figure 4-1. TWM process with shared planning

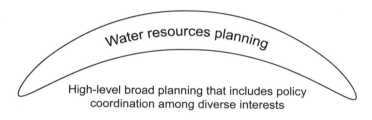

Middle-level planning that focuses on coordination between projects and purposes

Tactical-level planning that focuses on a specific project

Figure 4-2. Planning by levels

Planning and shared governance in TWM

In the TWM definition, three attributes relate directly to planning and shared governance:

- Conduct planning and management on a dynamic basis that adapts to changing conditions and local and regional variations;
- Seek participation of all units of government and stakeholders; and
- Effect decision-making through a process of coordination and conflict resolution.

These are not just buzzwords. They introduce important concepts

Table 4-1. Attributes of a water resources planning process

Attribute	Why attribute is important for water planning and management
Dynamic	Process stays in motion and achieves results, not static
Adaptable	Adapts to changing conditions and local and regional variations
Participatory	Participation of all units of government and stakeholders
Balanced	Uses coordination and conflict resolution to achieve fair results that consider social values, cost effectiveness, and environmental benefits and costs; and foster public health, safety, and community goodwill

that have evolved over the years to define a water resources planning process that serves the needs of stakeholders. Table 4-1 details the attributes of the process. Looking at these concepts a little further explains why they are important.

The expanded definition of TWM that was presented in chapter 3 adds further requirements to planning:
- Integration of management of water and land;
- Recognition of the role of law and regulations in TWM; and
- Promoting practices that enhance ecosystems as well as water.

As we have already seen, however, formal planning and management mechanisms are necessary but not sufficient for Total Water Management. To handle the myriad small issues that fly under the regulatory and management radar screens requires emphasis on citizen actions to go along with formal organizational actions. The other attributes in the expanded TWM definition that address these actions aim at its citizenship facets and should be considered in policy planning. They include:
- Stewardship by all sectors of society; and
- Going beyond compliance with laws and regulations.

Water resources planning

TWM is a way to couple the planning process with shared governance to reach decisions that add to the Triple Bottom Line. Therefore, good practice of water resources planning should lead to TWM.

Evolution of planning as a discipline

Along with the TWM and Integrated Water Resources Management (IWRM) concepts, the field of water resources planning has evolved over the past hundred years. Prior to about 1900, little government planning occurred in the water field. This era featured poor water management and urban sanitation; environmentalism had not taken root on this level in a significant way. What is known today as planning was not on anyone's agenda, and managers were just beginning to develop methods like economic analysis.

After 1900, with electric power being new and after several devastating floods, the government began planning for flood protection, hydropower, and navigation. Leading thinkers could see that these objectives were related to each other. By the time the United States entered World War I, the concept of multipurpose basin development was in use. The Corps of Engineers was receiving money and power to construct dams under the flood control acts of that time. During the 1930s New Deal era, the Tennessee Valley Authority was created during Roosevelt's ambitious "Hundred Days" period as a multipurpose river planning agency. Then, under the work of the National Resources Planning Board, the government developed many plans for river basins. As evidence that central planning is easy but implementation is hard, these masterpieces of planning work are mostly gathering dust in library archives today.

> Planning under TWM: dynamic, adaptable, participatory, balanced

After World War II, water planning received new impetus through work of the 1950s Senate Select Committee on Water Resources. In this postwar environment, emphasis was on practical problems such as building new dams and pipelines and providing housing for a growing country. Again, leading thinkers, drawing from their experiences during the New Deal, developed concepts for the Water Resources Planning Act (WRPA) of 1965, which provided for (see Figure 4-3):

- Establishment of a government Water Resources Council;
- Establishment of principles and standards for water resources plans;
- Periodic national water assessments;
- Financial support for state planning programs;
- River basin studies; and
- River basin commissions.

The WRPA introduced programs to provide a multilevel approach to water planning that coordinates among stakeholder interests. Congress thought a national framework such as this would work, but the programs of the act were terminated in 1981 during the Reagan Administration, which thought the act relied too heavily on government solutions. The WRPA did feature strong federal involvement, and then president Ronald Reagan thought that programs like this were the last gasp of the New Deal.

98 TOTAL WATER MANAGEMENT

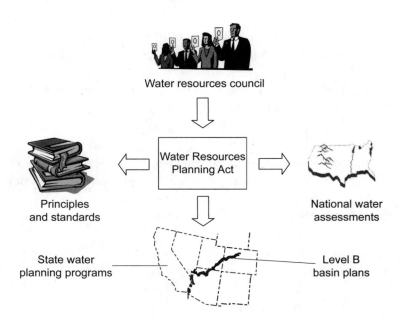

Figure 4-3.   Features of the Water Resources Planning Act

Terminating the programs of the WRPA was a shock to those who believed in centralized government-led water planning, and it left no structure in place for formal coordination of water planning. The water industry was left with a patchwork of laws and other institutional arrangements with which to work out coordinated plans. Figure 4-4 shows how water resources planning had evolved as of this time.

While the WRPA's programs were being developed, civil engineers recognized that more attention to planning and policy was needed, and in 1973 they created the Water Resources Planning and Management Division of the American Society of Civil Engineers (ASCE)[1]. This program, now merged into ASCE's Environment and Water Resources Institute (EWRI), provides a forum to discuss interdisciplinary topics of water resources (Committee on Water Resources Planning, 1962).

AWWA (2001) also recognizes water resources planning as a discipline and has a committee and a manual of practice about it, M50, *Water Resources Planning* (second edition). The manual focuses on how utilities can develop integrated resource plans with an emphasis on estimating

---

1 Maurice Albertson and Victor Koelzer, professors at Colorado State University, were leaders in this effort. Koelzer had worked on the staff of the National Water Commission with Ted Schad, who was the staff coordinator for the Senate Select Committee on Water Resources.

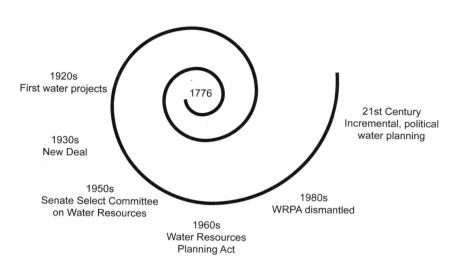

Figure 4-4.   Phases of the water resources planning act

demands, evaluating new sources, involving stakeholders, and responding to environmental regulations. As explained in chapter 3, integrated resource planning (IRP) focuses on new supply development, whereas TWM is a broader process that deals with stewardship, planning and management, and balancing competing uses through efficient allocation. In that sense, IRP is for specific projects, but comprehensive planning is a broader concept. For example, a river basin plan that involves multiple stakeholders is broader than the IRP for a single utility. IRP would be at the lowest level of Figure 4-2.

Now, after 30 years of experience with water resources planning and management, and after experience with planning in other disciplines, experts have noted how silos develop between the disciplines. Silos refer to the walls of separation between departments in an organization, where people don't talk to each other. This creates difficulty in *integration* of different points of view in all kinds of planning. Integration is the main purpose of TWM, which "balances competing uses of water . . . through a process of coordination and conflict resolution . . . and fosters . . . community goodwill." That, of course, is a tall order because of the high level of competition and conflict in some water decisions.

IRP focuses on supply development, TWM is for the "big picture"

For example, disciplinary stovepipes define the areas of *modeling* and of the *politics of water*. This means that one camp of specialists focuses on quantitative methods that lead to models for quantitative decision-making, while the other focuses on the soft side, which we simplify by naming the politics of water, a term that draws in a range of social and government issues. Each of these areas involves wide bands of topics, of course, and there are instances where the modelers work well with the policy scientists.

Planning process

The beginning point for planning is to create a rational problem-solving sequence that operates within administrative law and the political process. There are many ways to illustrate this planning process, but in some ways it is like the blind men who find an elephant. Each one feels a different part of it and thinks it is something different.

Theorists try to define terms and give some coherence to the discussion. For example, Branch (1998) wrote that government, business, and military planning follow the same process. He presented a basic definition for planning: "the process of directing human activities and natural forces with reference to the future." The problem with this definition, and others like it, is that you more want to include, the more abstract it sounds.

At its core, planning involves a problem-solving sequence to address this question: For a given goal for water management, what is the best way to accomplish it? The steps are:
- Problem identification
- Goal-setting
- Assembly of information on options
- Evaluation of options
- Decision-making
- Implementation
- The operations and control phase

In one form or another, this sequence underlies every formal water resources planning process. In the real world, it is a technical process that operates inside a political environment and bridges the gap between what are called the *rational* and *political* models of planning.

In the technical process, you study the problem, develop alternative solutions, weigh them, prepare analyses and reports, etc. In the political process, you consider interest-group and stakeholder involvement in decisions.

Few planning problems can avoid the political dimension. Even a simple example, such as establishing the safe yield of a reservoir within defined limits of drought risk tolerance, might affect stakeholders and have financial implications that lead to political controversy. An example of a rational problem inside a political environment might be a regulatory decision about a water quality standard that has costs and benefits disputed by different scientists and that might disrupt a local economy. In this case the stakeholders might array themselves on one side or other of the issue, depending on their perceived gains and losses according to the outcome.

The political process considers the divergent agendas of the players and requires identification of the players and interest groups, consideration of trade-offs and negotiating strategies, public participation, and establishment of incremental alternatives, as well as far-reaching solutions, consideration of individual and group preferences, analysis of voting behavior, and other political concepts. Figure 4-5 shows how the rational planning process works inside of a political environment.

> **At its root, the planning process is for problem-solving**

Along the route to a decision lie crucial decision subpoints that involve some or all of the stakeholders. These can be meetings, reviews, completion of studies, permits, new developments, and other events. During the process, influence and power are shifting, and knowledge is building. There may be a lack of organized information and intelligence about what is going on with allies, neutral parties, and opponents in water resources problem-solving.

Administrative law places boundaries around the political planning process. It requires hearings, permits, approvals, findings, and other decisions required by law.

Within these broad outlines, water managers have learned that there is an *art of planning*. These points capture some lessons that frame this art (Schmit, 2004):

- Planning has disincentives: it takes time, seems boring, may introduce contention into community discussions, and is expensive;
- Planning has benefits: funders expect a plan; it brings people together, provides a comprehensive view, anticipates regulatory challenges, and compares local experience with national environmental planning; and
- Do planning well: take time to do it right, encompass all values of the watershed, learn from someone else, recognize there is no cookbook solution, create your own plan, don't forget prevention and protection, and balance planning and doing.

Principles and guidelines for planning

Back in the 1960s, policy makers thought you could standardize aspects of planning, and under the Water Resources Planning Act of 1965, the US Water Resources Council (1973) prepared "Water and Related Land Resources: Principles and Standards for Planning." While this occurred

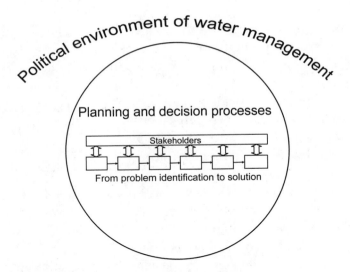

Figure 4-5. Rational planning in a political environment

more than 30 years ago and planning is very hard to standardize, these Principles and Standards (P&S) in their present form still offer useful guidance.

To signal that they were not standards, the P&S were downgraded in the 1980s by the government to *Principles and Guidelines* (P&G). The P&G offer two basic accounts: national economic development (NED) and environmental quality (EQ). Social well-being (SWB) accounts were relegated to a different category, Other beneficial and adverse effects, which were also to be displayed in plans. Regional development (RD) accounts were also included in this category. In the first drafts of the P&S, SWB and RD received parallel treatment to NED and EQ, but they were downgraded in favor of the economic and environmental objectives.

> **To deal with the political aspects of water you practice the "art" of planning**

When the P&S were downgraded to the P&G, mention of the SWB and RD accounts was omitted, but the information about the SWB accounts in the earlier documents remain valuable for use in social impact assessment (Economic and Environmental Principles and Guidelines, 1983). Chapter 8, which explains social impact analysis, draws from these earlier versions of the P&S.

Categories of beneficial and adverse effects listed in the P&S illustrate what they expected from water resources development. Table 4-2 outlines these effects.

Governance and shared governance

TWM requires shared governance because it "requires the participation of all units of government and stakeholders in decision-making through a process of coordination and conflict resolution." So the process of coordination and conflict resolution requires the sharing of power and decisions and is the central aim of the field of water resources planning.

The paradigm of shared governance explains how decision-making should occur among units of government and stakeholders with different interests, and TWM also explains the criteria for the decisions in that they should balance "competing uses of water through efficient allocation that addresses social values, cost-effectiveness, and environmental benefits and costs." This is in effect a statement of the Triple Bottom Line of economic, social, and environmental values.

Governance in water management

To understand shared governance, we begin with the concept of governance itself. Given its shared nature, management of water requires some involvement of government, especially for regulation and coordinating mechanisms. Given the drag caused by government corruption and incompetence, leading thinkers have identified governance as a necessary condition for effective water resources management.

Governance is the act or process of governing and has meaning close to that of politics, when politics means the art of government. You might say that politics is the theory and governance is the practice of government.

Because politics is the art of government, the political process is important for government to function, but also a way to negotiate conflicts and balance outcomes to meet goals and objectives within the economic framework. Water decisions have a high political content because people have different agendas that should be worked out in a political process. Politics and government provide people with rules and processes to resolve their differences and make positive things happen.

Shared governance in water planning

If governance is a challenge, then shared governance is even more difficult. *Shared governance* is the collaborative sharing of authority and resources in the management of societal institutions. Sometimes it is difficult to have balanced and positive outcomes for all stakeholders among the competing agendas. The principles of shared governance help establish processes that end up with the required balance and positive net benefits.

Shared governance is sharing of management of societal institutions

Table 4-2. Beneficial and adverse effects of water resources development by category*

Category	Effects
National Economic Development (NED)	• Value of increased outputs of goods and services (+) • Value of resources used in a plan (-) • Value of output from external economies (+) • Losses in output from external diseconomies (-)
Environmental Quality (EQ)	• Open and green space, lakes, beaches, shores, mountains, wilderness areas, estuaries, wild and scenic rivers, and other areas of natural beauty (+ for gains, - for losses) • Archeological, historical, biological, and geological resources and selected ecological systems (+ for gains, - for losses) • Quality of water, land, and air resources (+ for gains, - for losses) • Irreversible commitments of resources to future uses (-)
Regional Development (RD)	• Increased outputs of goods and services in regions (+) • Outputs in regions from external economies (+) • Value of regional resources used in a plan (-) • Losses in regional output from external diseconomies (-) • Jobs (+ for gains, - for losses) • Population distribution (+ for gains, - for losses) • Regional economic base and stability (+ for gains, - for losses) • Regional environment (+ for gains, - for losses) • Impact on regional development goals (+ for gains, - for losses)
Social Well-Being (SWB)	• Real income distribution (+ for gains, - for losses) • Life, health, and safety (+ for gains, - for losses) • Education, culture, and recreation (+ for gains, - for losses) • Emergency preparedness (+ for gains, - for losses) • Other social effects (+ for gains, - for losses)

Source: Economic and Environmental Principles and Guidelines, 1983.

* The + and − symbols refer to beneficial and adverse effects, respectively.

Shared governance is familiar concept in business-labor relations, where managers collaborate with workers in reaching decisions, whether in unionized or nonunionized environments. Another nongovernmental example is shared governance in universities, where professors work with administrators to reach decisions.

Examples of shared governance in water decisions would focus on regional cooperation and intergovernmental coordination, two important but difficult goals. A shared effort also means having public support for the shared goals. It will involve units of government and stakeholders in coordination and conflict resolution to develop water plans and projects such as water supply sharing, regional treatment plants, regional reporting and coordination of water uses.

It should be emphasized that, while utilities have the core responsibility, all citizens in a democratic society have some role in water resources planning. The challenge with water is how, with a complex system, to give

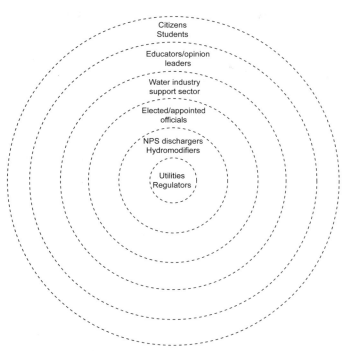

Figure 4-6. Circles of responsibility for water planning

everyone a say but still take care of business. Figure 4-6 shows the circle of responsibility for water resources that explains shared responsibility.

Examples of decision problems requiring planning and shared governance

Water resources decisions requiring planning and shared governance can include any issues that involve facilities planning, operations, finance, or regulatory control (Grigg, 2005). The most visible decisions might be to build or renew a facility, obtain water supplies, or develop new capacity for water services. Facilities might include dams, water or wastewater treatment plants, or pipelines, all of which can be shared. Water supplies can involve either access to new or increased sources of water supply. Joint planning for these might result in shared facilities or simply coordination of development of separate facilities on a win-win basis.

Operating decisions might be to alter the release schedule from a reservoir or change wastewater treatment methods, for example. If the facilities are shared, the decisions require shared governance, and if not, they require coordination among those

Water resources decisions focus on planning, operations, finance, and regulatory control

impacted. Emergency planning and security are good opportunities for joint work. They can include to develop a multipronged plan to respond to flood, drought, or other threats, including mitigation, response, and recovery, with mutual aid.

Regulatory decisions by their very nature usually require shared governance. For example, rule-setting has broad impacts, and issuing a permit or license has long-term and broad effects.

Financial decisions, such as to provide funding for facilities development, open a window of opportunity for shared planning and governance. If economies of scale can be gained by collaboration or if funding responsibilities can be spread across more taxpayers or ratepayers, risk can be reduced for all concerned.

Defining roles and relationships

Many different groups participate in the planning process. The case study in chapter 2 identified groups that would be involved in a regional decision, and the chapter also outlined how the water industry is organized. Using this model of organization, Table 4-3 lists typical players and roles in the TWM processes. In addition to the direct roles, all of the groups can undertake their corporate social responsibilities and perform outreach, as well as practice conservation and encourage others to do so.

Integration and coordination through shared governance

Coordination and *integration* are the main tasks of shared governance. Integration means to blend things together and coordination means to create order and harmony among them. The requirements for them include stakeholder involvement; geographic integration through basin plans; communication systems; permits, licenses, and rights; negotiation and conflict resolution; intergovernmental coordination; functional coordination; and coordination among knowledge areas.

As the Internet leads to direct democracy, stakeholder involvement requires more sharing of information

Stakeholder involvement

Stakeholder involvement in water planning is a democratic process that requires fair and transparent governance, open decisions, sufficient notice, and the opportunity to engage in decisions. Driven by the Internet and the news media, the United States is moving toward more direct democracy

Table 4-3. Players in the planning processes

Category	Players	Direct Roles
Water service providers	Water supply directors (cities) Wastewater directors Irrigation company boards and staff Stormwater utility directors Electric power generators	Perform missions and serve customers
Regulators and other government agencies	Federal regulators State EPA directors US Army Corps of Engineers executives (civilian or military) USGS District Chiefs State water resources directors	Perform missions; regulate effectively
Support sector	Consulting engineers Water and environmental attorneys Think tank representatives Environmentalists Vendors	Provide range of services and products
Water users and impacters	Land developers Large scale farms Groundwater pumpers Water recreation representatives Land or stream modifiers Nonpoint source dischargers Public at large	Use water responsibly; comply with and exceed regulations; practice stewardship

with more chances to participate and vote on things than in the past. To participate effectively, people must understand the issues.

The planning process must consider the divergent agendas of the players, so it is important to identify them. There will always be people who seek to drive the agendas, and some people will always be leaders and others followers.

The mechanics of stakeholder involvement require management programs staffed by competent staff. Agendas are set for plans and actions so citizens can participate. It is important for the process to use people's time efficiently.

Geographic integration through basin plans

Usually we think of a watershed or river basin as a geographic unit where balance is required. Within these, water users compete with each other to divert water, build storage reservoirs, generate hydroelectricity, use water for recreation, and even avoid water regulations that might cost them money.

A watershed could be small, say a few square miles, and still have competing users. Or a river basin could drain a whole state and require coordination among thousands of water users. So the watershed or river basin is the main unit for planning the balance among water users.

Communication systems

Communication systems are required for coordinated planning. They require meetings, web pages, shared-vision model sessions, and other venues where the stakeholders share information among themselves.[2] Reporting and information systems are important vehicles for communication.

Permits, licenses, rights

Applications for permits, licenses, and rights to water set planning processes in motion. These decisions form integral parts of the planning process because they focus the attention of the stakeholder on one set of requirements or another. Once an authority has a permit and an infrastructure, follow-up is required to ensure that the conditions assumed during the planning phase hold up during the operational phase.

Negotiation and conflict resolution

Perfect agreement among stakeholders is usually not achieved, and shades of meaning have developed for the term consensus. Levels of *consensus* can include:

1. There is a high level of disagreement.
2. There is enough disagreement to indicate the decision will not work.
3. The disagreement and agreement are about the same.
4. While all stakeholders do not agree, to move ahead they can live with the decision.
5. All stakeholders agree with all aspects of a decision.

A pragmatic approach might have the stakeholders agreeing from the onset that if level four is reached, the decision can proceed.

Intergovernmental coordination

Intergovernmental coordination means working in harmony between the levels of government (vertical coordination) and with other utilities or units of government at the same level (horizontal coordination). This is

---

[2] Shared-vision planning is a concept developed to forge consensus among participants by working together on modeling and studies in advance of decisions.

where a lot of the heavy lifting of shared governance gets done, by getting to know others, spending enough time together to learn about each other's problems, and deciding to work together through various means.

Functional coordination

Another type of coordination is across functions of management, such as engineers working with planners and with financial managers. It also means that operational managers work closely with staff managers. Generally speaking, it refers to having good communication and working relationships across the organization.

Coordination among knowledge areas

In planning, it is important that the knowledge areas and disciplines work together. You may have experts from several professions on a team, and they must respect each other and listen to other points of view. Scientists must work with engineers, and vice versa. Economists and lawyers must work with biologists. A specialist in health effects might need to work with someone who knows about hydrology.

Regionalization: its promises and challenges

Planning and shared governance in watersheds is a form of *regionalization* (Grigg, 1989). Conceptually, regionalization in water management could include integration or cooperation in a metro area or other geographical region as well. Regionalization is complex. A committee of AWWA (1981) presented this definition: regionalization is "a creation of an appropriate management or contractual administrative organization or a coordinated physical system plan of two or more community water systems in a geographical area for the purpose of utilizing common resources and facilities to their optimum advantage."

> **At the end of the day, TWM is all about coordination and sharing**

That definition focused on sharing of service delivery rather than planning, but the concept of regionalization goes further. It includes functional integration (as in water supply + wastewater), area-wide integration of any function (such as planning), and any form of cooperation to manage water.

Many policy studies have advocated regionalization to solve water industry problems. Advocates point out that it can provide benefits in economics, service, and water quality.

Barriers to regionalization are the politics of the issue, and they are formidable. These can include loss of control of income, the need for legislation, public indifference, distrust and provincialism, regional bureaucracy, added complexity, inequities in financing, personnel difficulties, public-private incompatibilities, and redistribution of revenues.

Possibilities for regionalization would include regional management authorities; consolidation of systems; a central system acting as raw water wholesaler; joint financing of facilities; coordination of service areas; interconnections for emergencies, and sharing of any management or service responsibility (including planning). Experiences with regionalization have shown that TWM's need for shared governance can expect the same promises and perils.

Toward the future

This chapter discussed how shared governance is essential in TWM to defeat the "it's not my problem" syndrome, and the way it gets done is through the planning process.

While our government system has institutional arrangements for shared problem-solving, it finds them difficult to work out.

If all goes well, it should result in roundtable-type activities, such as shown in Figure 4-7. Professionals from different jurisdictions should be able to get together and work things out.

Source: Alan Skrepnek, Manitoba Water Stewardship Department

Figure 4-7. Roundtable

Water resources planning, which is almost a discipline itself, offers paths to cooperation and joint problem-solving that support shared governance. Planning methods have been evolving for over 100 years, leading to today's decentralized process, which uses a number of sophisticated tools. This process is, in effect, a rational and linear process that operates inside of a political and legal environment.

Within TWM, the paradigm of shared governance explains how planning and decision-making should occur in ways that recognize sustainability and the Triple Bottom Line of economic, social, and environmental values. Shared governance requires recognition of the roles of many different groups, including water service providers, regulators and other government agencies, the support sector, and water users and impacters.

Summary points

- The TWM principles that relate to planning and/or governance include: to develop an effective TWM process, organize shared governance, plan for sustainable development on a watershed basis, set shared goals, and commit to coordination.
- Governance is the act or process of governing and has a meaning close to that of politics, when politics means the art of government. Shared governance in TWM is a mechanism for cooperation and sharing of power and decision-making, or a way to work together to solve common problems. Organizing shared efforts can be difficult because of institutional constraints and perverse incentives. Institutional constraints include organizations, cultures, laws and regulations, and incentives. Perverse incentives are those that create actions that are opposite of those that are needed.
- Shared governance requires participation of all units of government and stakeholders, and decision-making through a process of coordination and conflict resolution, two key principles of TWM.
- In water resources planning, TWM seeks integration of management of water and land, recognition of the role of law and regulations in TWM, and the promotion of practices that enhance ecosystems as well as water.
- Along with TWM and IWRM concepts, the field of water resources planning has evolved to focus on multidisciplinary and decentralized approaches to problem-solving. AWWA's manual of practice, M50, *Water Resources Planning*, focuses on how utilities can develop integrated resource plans.
- Water resources plans include political dimensions. Even simple issues affect stakeholders and require consultation and conflict resolution,

which require a rational planning process that works inside of a political environment.
- The "Water and Related Land Resources: Principles and Standards for Planning," originally prepared by the US Water Resources Council, define the state of practice of certain kinds of planning. These address national economic development and environmental quality. Social well-being and regional development accounts were downgraded in favor of economic and environmental objectives.
- Coordination and integration are the main tasks of shared governance. Integration means to blend things together, and coordination means to create order and harmony among them. These require stakeholder involvement; geographic integration through basin plans; communication systems; permits, licenses, and rights; negotiation and conflict resolution; intergovernmental coordination; functional coordination; and coordination among knowledge areas.

Review questions

1. Define governance and politics as they relate to water resources management.

2. Explain why shared governance in water management is difficult to achieve. Give examples.

3. A principle of TWM is that "Shared governance requires participation of all units of government and stakeholders and decision-making through a process of coordination and conflict resolution." Explain how this relates to the concept of Integrated Resource Planning.

4. Give examples of benefits you would expect to credit to water plans in the categories of national economic development, environmental quality, and social well-being.

5. Differentiate between the concepts of comprehensive planning, integrated planning, and coordination in water planning. Do these concepts illustrate nuances between important elements of the water resources planning process? Why or why not?

References

American Water Works Association. 1981. Regionalization: Why and How. *Jour. AWWA* 73 no. 5 (May).

AWWA. 2001. *Water Resources Planning.* 2nd ed. Manual M50. Denver, Colo.: AWWA.

Branch, Melville C. 1998. *Comprehensive Planning for the 21st Century: General Theory and Principles.* Westport, Conn.: Praeger.

Clawson, Marion. 1981. *New Deal Planning: The National Resources Planning Board.* Baltimore, Md.: Johns Hopkins University Press.

Committee on Water Resources Planning. 1962. Basic Considerations in Water Resources Planning. American Society of Civil Engineers, *Jour. Hydraulics Division* HY 5 (September).

Economic and Environmental Principles and Guidelines for Water and Related Land Resources Implementation Studies. 1983. Washington, D.C.: Superintendent of Documents.

Grigg, Neil S. 1989. Regionalization in Water Supply Systems: Status and Needs. American Society of Civil Engineers, *Jour. Water Resources Planning and Management Division* (May).

Grigg, Neil S. 2005. *Water Manager's Handbook.* Fort Collins, Colo.: Aquamedia Publications.

Schmit, Ayn. 2004. Comments on Water Resources Planning. Big Thompson Water Forum Annual Meeting, Loveland, Colorado, February. Unpublished.

Senge, Peter M. 1990. *The Fifth Discipline: The Art and Practice of The Learning Organization.* New York: Doubleday Currency.

US Water Resources Council. 1973. Water and Related Land Resources: Principles and Standards for Planning. Part III. *Federal Register* 38 no. 174: 24778–24869.

United Nations Development Program. 2006. *Lessons Re: Effective Coordination Mechanisms.* http://europeandcis.undp.org/WaterWiki/index.php/Coordination_mechanisms. Accessed September 28, 2006.

CHAPTER 5

TRIPLE BOTTOM LINE REPORTING FOR WATER AGENCIES

Without a scorecard, TWM can be just a visionary concept. With its focus on economic, social, and environmental benefits and costs, TWM lends itself to what is called Triple Bottom Line (TBL) reporting. In some ways, TBL reports might be just adding environmental and community information to a regular financial and performance report. Looked at another way, TBL reporting can add a whole new dimension to a utility's understanding of its broader responsibilities under TWM. This chapter is about how TBL reporting can facilitate the TWM process, both for individual utilities and for groups working together in planning and shared governance.

Think of a school without report cards or a ball game without a scorekeeper. Neither would yield the intended results because scorekeeping provides the incentive to perform. Scorecards and their indicators of success are used throughout life to track performance. Businesses report their financial performances, politicians track their approval ratings, and meteorologists vie to predict the season's hurricanes most correctly.

Given its goals of stewardship and balanced water management, the TWM process should include reports on how TWM affects large- and small-scale impacts on water resources. TBL reports include information on all three of these accounts, whereas a business report will focus on financial reporting only (Figure 5-1).

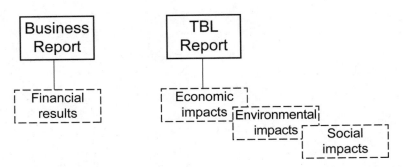

Figure 5-1. Business reports versus TBL reports

A TBL report is a display of achievements and setbacks in economic, environmental, and social categories. The economic accounts can include financial results, but they might also address other issues, such as economic development. Environmental and social accounts address positives and negatives for habitat, society, and related impacts. A TBL report can range from a regular financial report, augmented by economic, social, and environmental results, to a special, focused TBL report that focuses only on the economic, social, and environmental aspects.

A TBL report is not a panacea for all reporting. For one thing, a TBL report for a local water supply utility in the East might look a lot different from one prepared by a multipurpose water district in California, for example. Also, if it is not prepared well, a TBL report might divert attention from important business issues and not be considered as relevant to the central purposes of the organization. In spite of these factors, a TBL report can go a long way to address reporting issues of water agencies and, if used properly, can help agencies to identify and pursue their broader agendas as well as their core business lines.

A TBL report is not a panacea

The TBL report can help overcome the "it's not my problem" syndrome as well as the bigger issue of the balance between a utility's direct responsibility to customers and its external responsibilities to society and the environment. For example, a utility might deliver clean and safe water to its customers, but if providing those services takes water from another group of people, the result may violate social, environmental, and economic sustainability.

It is important to know who will use the reports and for what purposes. A reporting system without a purpose will not be sustained. The reports should be useful both to the players in the water industry and the public at large who can act on them. At the local level, the reports can provide an outlet for utilities to inform stakeholders about their environmental and social performance. TBL reporting can provide useful information about a utility's goals, outcomes, and impacts. Once you go outside the organizational boundaries of utilities, reports would have to be aggregated to show the big picture of value added by water management decisions. At the regional and broader levels, they could give the public the big picture of how the nation is doing in its water management and report on the sustainability of water use, based on results of monitoring and assessment of water use, water quantity, and water quality.

Scientific monitoring and assessment is expensive and complex, but citizens can be effective monitors, too. Water issues occur in neighborhoods and in rural areas, and boaters, fishers, and hunters can report about threats to waters if they are trained to do so.

TBL as sustainability reporting

Although TBL reporting received its name from the sustainability movement, for a long time the water management community has been aware of the need to report economic, social, and environmental impacts. Perhaps the main difference is that broad economic, social, and environmental reports were in the past thought of as planning reports, whereas the TBL report can be an augmented business report for a utility or agency.

Sustainability reporting might be thought of as a way to improve corporate governance and transparency in the area of environmental impacts. The TBL approach is receiving wide support among international groups such as the Global Reporting Initiative (2006). The Global Reporting Initiative began in 1997 as an idea for a disclosure framework for sustainability reporting. The initiative came from CERES (2006), a Boston-based nonprofit that was formed in 1989 as a partnership between environmental groups and institutional investors.

TBL reporting came from the sustainable development movement

Each organization would develop its own unique approach to TBL reporting, perhaps using guidelines such as those by GRI. GRI has a reporting framework with protocols for economic, environmental, and social accounts, and also for human rights, labor, and product responsibility. The TBL is a report for utility performance across the board. Specific examples are given later of utility TBL reporting.

TBL as multicriteria scorekeeping

By definition, TWM involves progress in broad areas that include the economy, environment, and society. It is like keeping up with collective statistics in several games at the same time, which taken together score how well sports are going, rather than how well with just a single game like football.

The broad goals of TWM and its focus on sustainability suggest that the TBL can be used to keep score. Other scorecards can also be used. In particular, the *balanced scorecard* (BSC) and *multicriteria decision analysis* (MCDA) can be used.

MCDA

The Triple Bottom Line requires reports of economic, social, and environmental gains and losses. The tool of multicriteria decision analysis provides a framework to organize evaluation information for these categories of goals and to study their trade-offs. It evolved from tools of economic, environmental, and social impact analysis and stems from welfare economics and utility theory.

> **Planning under TWM: dynamic, adaptable, participatory, balanced**

While MCDA seems straightforward, there are many ways to display its results. You may have numeric data combined with qualitative data on people's preferences about choices. Given that it is not precise, an MCDA exercise is a way to display information, not the final word about a range of choices. Decision makers should study the information and carefully consider its sensitivities to assumptions.

In its simplest form, an MCDA display shows how strategies or projects score in the goal categories, as shown in Table 5-1. In the display, you provide a net score or verbiage about each project in each category. To do this, you must be able to evaluate the projects to determine the scores, and you must have a scoring system.

Table 5-1. Scoring strategies by goal

Project	Economic	Environmental	Social
Project A	(scores here)		
Project B			
Project C			

A good example of MCDA reporting is in a report about how Tampa Bay Water in Florida should manage its multisource water system (Tampa Bay Water and CH2M Hill, 2006). This MCDA report is embedded in a model that shows the utility how to select water from its different aquifer and surface water systems while considering cost minimization, environmental stewardship, and source reliability. MCDA reporting is a way to show performance or scores in different categories, and MCDA and TBL reporting have similar purposes.

Balanced scorecard

The balanced scorecard is another tool from the business world that is related to the TBL in the sense that it seeks to measure more than a single economic or financial indicator. The BSC concept originated at Harvard and is credited to Kaplan and Norton (1996). It has become popular and been the subject of numerous articles and books such as Niven (2006).

BSC and TBL are similar, but they concentrate on different goals

As applied to business, the BSC measures four areas of outcomes: financial results, customer relations, internal business processes, and learning and growth outcomes. The idea is that the health and performance of a business go beyond just the financial bottom line.

Financial results have a focus on the *bottom line*, with financial ratios, balances, and net sums to be reported. Customer relations measures how well you are doing with your customer base on an integrated basis. Internal business processes focus on the integrity of your infrastructure, on your procedures, on how well you tend to externalities such as environmental control, on your human resources management, on legal and risk management, and other processes. Learning and growth are an important measure of how well individuals advance and how well the organization learns and grows as a unit. Today's search for talent in organizations emphasizes the importance of individual capacity, and organizational capacity is a function of the sum of individual talent and how well workers perform as a unit[1].

Figure 5-2 shows how the BSC compares to the TBL report. The difference focuses on which areas of business operations each reporting system emphasizes. The BSC emphasizes a wider range of internal and customer issues, whereas the TBL emphasizes external impacts and sustainability more. For this reason, and given the heavy impact of water management on the environment, TBL reporting is better for water utility reports to stakeholders and the public, whereas a BSC approach might be

---

[1] *The Economist* magazine had a special report on the global search for talent in its October 7, 2006, edition (Battle for Brainpower, 2006).

better for reporting an integrated business assessment, say for the utility's board of directors.

If applied to a water utility, the BSC would mainly address how it operates as a business, and this would have to include how it impacts the environment and society. However, the BSC has been developed with business survivability and growth in mind and is not set up for TBL reporting.

The TBL's financial and economic part would measure debt, rates, reserves, and whether the utility is the best of class in finance. In the social category it might consider how a utility can have impacts on social variables such as poverty and health. In the environmental arena, it could look at which part of the environment the utility can impact and measure the gap between ideal and actual. TBL might be used here as an input to the utility's overall BSC, with other performance measures used for the rest of the BSC[2].

Goals from the definition of TWM that might be considered in a BSC report are community goodwill, stewardship, the greatest good of society and the environment, sustainability, public health, safety, social values, cost-effectiveness, and environmental benefits and costs.

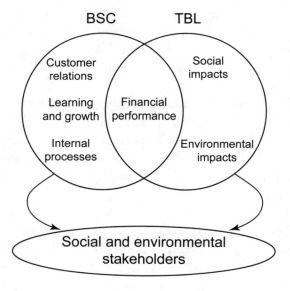

Figure 5-2.   TBL and the Balanced Scorecard

---

[2] See Alegre et al. (2006) for a comprehensive report on performance measurement in the water industry.

The BSC and TBL frameworks are very similar, both in their rationale and in their presentation. The BSC offers four accounts, whereas the TBL has three accounts. The BSC has tools, such as a strategy map, that can be used to improve planning in the three areas of the TBL.

Use of indicators in TBL reports

Reporting involves art and science. The science is in compiling the numbers and displays, and the art is in deciding what is important and how to present the information to have the best effect. In sports, business, or even politics, scores track how you are doing toward achieving your goals. In business, scoring is by the financial bottom line and by stock prices, among other indicators, such as growth and return on investment. Business reporting is regulated by agencies such as the Securities and Exchange Commission (SEC) and the Internal Revenue Service (IRS). Companies compile and issue comprehensive annual financial reports (CAFR) so that investors and regulators can monitor their activities. Standards for reporting are also promulgated by other agencies, such as the Financial Accounting Standards Board (FASB).

Indicators are a way to summarize performance

In addition to their required financial reports, businesses are now reporting more about their corporate social responsibility (CSR)[3], as well as preparing balanced scorecards.

Keeping score in politics and government is more ambiguous. In politics, ratings include polls, elections, and enactment of legislation, among other things. In government, measures are introduced for efficiency and effectiveness but are not as visible to the public as business success is. Keeping score of the federal government is a function of the Government Accountability Office (GAO), for example, and the public only hears about its reports when they seem newsworthy.

On a relative basis, water utilities and agencies have more responsibility to report social and environmental impacts, while private businesses have more responsibility to report financial results. This is illustrated in Figure 5-3, which shows how corporations report to investors and to financial regulators, and how utilities report to customers, stakeholders, and governing boards.

Business reporting is not as simple as shown in the figure. Different types of public and private organizations have different reporting

---

3 CSR is explained in chapter 11.

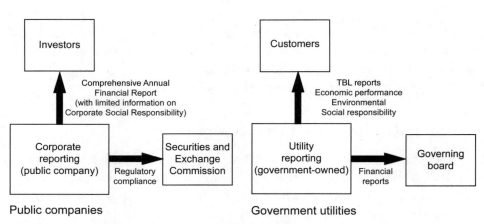

Figure 5-3. Reporting by public companies and government utilities

requirements. For example, some water utilities are investor-owned, regulated businesses, and these must report both financial profits and regulatory compliance in the water arena. Other businesses have to report on their regulatory compliance as well. For example, auto manufacturers must report on the safety and gasoline consumption of their products.

Measuring results with indicators

When preparing a TBL report card, careful attention must be paid to measuring both the results and the status of the water systems. Managers and the public require clear and credible information based on indicators that can be measured and reported objectively. A TBL report ought not to be just public relations but should reflect valid and objective information.

The options available for reporting of performance such as in the TBL format have expanded with new information about indicators. The more comprehensive the reporting format, the more important it is to have clear sets of indicators. Otherwise, the reporting goes all over the place.

Developing indicators for TWM

Indicators in the TBL report ought to be aligned with the utility's goals that relate to economic, environmental, and social accounts. In turn, if the utility is serious about these goals, then it faces decisions if it is underperforming in any of the goal categories.

Attributes of indicators

In designing the system of indicators, the paramount issue is that they communicate well to the stakeholders and public. A good bit of thinking has gone into designing systems of performance indicators. Table 5-2 describes attributes they should have.

Table 5-2. Preferable attributes of a system of indicators

Attribute of indicators	Explanation
Comprehensive	They should be comprehensive enough to cover all activities measured.
Clear	They should be clear so they can be understood without jargon.
Integrated across levels	They should provide aggregated information from lower to higher levels. Higher-level indicators should be based on lower-level data with integrity.
Aligned with decisions	Indicators should measure goals and decisions related to goals.
Uniform language	Uniform language should be used to communicate with a wide band of stakeholders.
Packed to increase density	The information should be packed to communicate as much information as possible with fewest words.

Source: Grigg and Vlachos, 2005.

The research indicates that indicators should communicate accurately and concisely with information that is relevant and useful. Figure 5-4 illustrates the challenge, as shown for environmental indicators. The indicators shown are just examples. In reality, the same challenge exists even for financial indicators, such as a debt ratio. Think of all the detailed

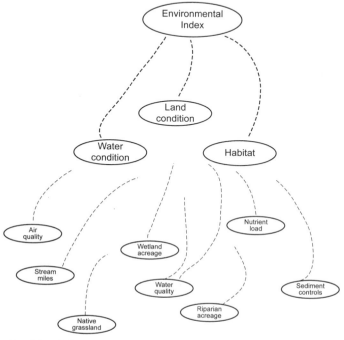

Figure 5-4. Packing information to create an environmental indicator

financial information that must be processed to come up with summary reports.

Water industry indicators

While it has good guidance on financial indicators, the water industry lacks consensus indicators for social and environmental accounts. Also, its financial indicators must be processed so that they communicate well to the public. AWWA has some performance indicators in its water industry database, and USEPA collects some data on the performance of water supply systems. AWWA's QualServe program includes indicators. In Europe, six cities developed a comprehensive indicator system for water utilities (Stahre, Adamsson, and Eriksson, 2000), and the European Union has sponsored research on performance indicators (Alegre et al., 2006). None of these are broad enough for TBL reporting, however.

Economic, environmental, and social reporting

In the economic category, a TBL report can provide just an overview of financial information, or it can contain the detail you would expect in comprehensive annual financial reports. Most water utilities issue an annual report for their customers and regulators, but for government-owned utilities, the reporting is not usually regulated other than that public finance follows rules of the Governmental Accounting Standards Board (GASB).

> A TBL report is in many ways an accounting report

Chapter 7 gives some detail on environmental accounting, which is a wide-open field without any real standards. Actually, the emergence of environmental indicators is meant to provide a language and structure to accounting and reporting. Chapter 8 explains the approaches to social impact accounting. These are less structured even than environmental reporting. This lack of structure for environmental and social accounting and reporting emphasizes the importance to the organization of creating goals to measure its achievement against and of the credibility of the TBL report, as opposed to being simply public relations.

TBL reporting for water management

The basic criteria for scoring TWM is given by its basic definition, that it is "the exercise of stewardship of water resources for the greatest good of society and the environment." It follows that the scorecard should measure the good of society and the environment. TBL reporting is aimed at measuring this "good" to include economic and financial results.

General theory

This aim is linked to a concept called *welfare economics*, which aims to explain the good of society by a social welfare function. The equation for the *social welfare function* is:

$$SWF = a_1 * X_1 + a_2 * X_2 + ... a_n * X_n;$$

where SWF is the social welfare function; X represents categories of public goods as measured by the size of the economic pie, more environmental quality, and social benefits; and a represents the weighting functions of how society values these public goods. In some ways, it is like the guns-versus-butter trade-off that economists like to cite for choices a government makes.

The definition of TWM does not tell us how to value these, but it does state that TWM "balances competing uses of water through efficient allocation that addresses social values, cost-effectiveness, and environmental benefits and costs." However, it does not tell us *how* to balance them or address them.

In applying the SWF to TWM, it is difficult to decide the weighting factors because people have different ideas about them. Resolving this is more in the realm of politics than in economics. That is the same issue as any multiobjective decision-making, which requires weighting factors among goals.

Thus, the TWM scorecard is a framework to display social values, cost-effectiveness, and environmental benefits and costs. In other words, it is a display of the economic, social, and environmental accounts like the ones called for by the Principles and Standards of the 1965 Water Resources Planning Act (WRPA). These provided a multiobjective display of project and program impacts for national economic development (NED), regional development (RD), environmental quality (EQ), and social well-being (SWB). Their history and fate are explained in chapter 4.

TBL accounts for social welfare in its broadest sense

Lest we wonder if there is anything new with TBL reporting, there is. What is new is a greater understanding of society's preferences for balance among these accounts, new frameworks to display information, and greater understanding about how society is going to make its choices among the alternatives.

Current status

A number of water industry leaders are now adopting TBL reporting to indicate that utilities have responsibilities beyond their own units. According to a recent research report from AwwaRF, TBL reporting is viewed by utilities as a way that they can report on their responsibilities to the environment and the community, as well as on financial matters (Kenway and Reekie, 2006).

Benefits to TBL reporting are cited in the report as:
- More trust among staff, customers, regulators, and stakeholders on sustainability issues;
- Creating reputation and brand;
- Attracting and retaining employees;
- Improving access to investment markets with socially responsible shares;
- Reducing risk profile by sharing information;
- Cost savings by rationalization of public reporting;
- Benchmarking and performance measurement of sustainability;
- Assisting improvement of strategies and plans; and
- Providing a sound basis for stakeholder dialogue.

Translating TWM goals into a scorecard

Although the concept of TBL is simple, it faces major problems. First, how do you do it, and how do you assure integrity in such a scorecard? Then, even if one utility had a good TBL score for its TWM activities, how do you score the collective efforts of all players with shared responsibilities?

TBL reports can build trust through transparency and accountability

The definition of TWM gives us a list of goals that help us to create a scorecard, but these goals are general and must be translated into specific measures that fit different situations. Table 5-3 details these goals and measures. Other goals are implied by the definition, such as to promote water for habitat and to improve water security.

Table 5-3. TWM goals and specific measures

TWM goal	Measures
General goal	Exercise of stewardship for the greatest good of society and the environment
Business, economic, or efficiency goals	Promote supply development Promote water conservation Promote reuse Promote source protection
Environmental and water quality goals	Enhance water quality and quantity
Social goals	Foster public health and safety Foster community goodwill

While these form a general guide, they are not specific enough for a given utility or agency. They offer a framework for TBL reporting, but each utility, agency, or organization must develop its own scorecard based on its unique mission and situation.

Status of TBL reporting in the water industry

To water planners, the concept of TBL reporting has been evolving for a long time. The need for national water assessments to consider economic, environmental, and social impacts was recognized 40 years ago in the 1965 Water Resources Planning Act (WRPA) and later in the Clean Water Act. Under the WRPA, the Water Resources Council compiled National Water Assessments, but the program was halted some 27 years ago. Later in the 1980s, the USGS prepared National Water Summaries, but the report series has not been continued. In other words, the idea of TBL was around, but it was not enabled and is not currently in use for national-level water planning.

> **There are many water reports, and TBL is a way to integrate them**

Just as planning occurs at several levels, reporting can as well. At local levels, very few utilities do TBL reporting. Compared to the relatively few large, regional utilities, most of the predominantly small utilities in the United States would not see themselves as having comprehensive purviews. Their reporting will focus on legal mandates such as the Consumer Confidence Report (CCR) under the Safe Drinking Water Act (SDWA). Wastewater utilities would also file required reports under their permit requirements, but at the local level, no entities are mandated to do TBL reports covering economic, environmental, and social impacts.

At the national level, the Clean Water Act provides for biennial national water quality summaries through the Section 305(b) process, but their scope is limited to a compilation of state reports. The national Section 305(b) report is to contain information on the condition of all waters in the states and information on pollutants (chemicals, sediments, nutrients, metals, temperature, pH) and other stressors (altered flows, modification of the stream channel, introduction of exotic invasive species) that impair waterbodies (USEPA Watershed Academy, 2006). The 305(b) report gives a partial picture of the condition of the nation's waters, but it lacks much of the information and assessment required to present a comprehensive report. Right now, only nongovernmental organizations compile comprehensive report cards on the nation's water, and these may reflect the biases of the issuing organizations and not be based on thorough and objective analysis.

At the regional level, councils of government and similar entities might prepare *environmental reports*. These will tend to be very general and not focus on the impacts of water management actions. The same would be true for the state and national levels. At the state level, departments of environment and/or natural resources prepare reports and have information on their web pages. Nationally, the Council on Environmental Quality is empowered through the National Environmental Policy Act (NEPA) to prepare reports, but their reports seem to be compilations of reports that others have issued. USEPA also issues comprehensive environmental reports. However, no national agency prepares a TBL report that focuses on water actions. As a result, reporting is fragmented.

Utility TBL reports

An individual utility or similar organization can describe its accomplishments in the three categories. This description would be much like a corporation's explanation of its overall record over the past year in a balanced scorecard format.

There is no set format for utility TBL reports, but one way to do it would be to start with a strategic planning exercise to identify the economic, environmental, and social issues that the utility can address. Once the plan is ready, a comprehensive TBL report can be prepared on a one-time basis, or at least one that will last a few years. Then updates can be placed in the utility's annual reports. Alternatively, the utility can include more comprehensive TBL information with its annual financial reports.

Some examples of issues that might be included are described in Table 5-4. Reporting TBL results for a given organization will cover its direct responsibility. If it also includes its corporate social responsibility, then it "goes the extra mile."

As a starting point to identifying goals and indicators, the definition of TWM might be used. Taking key elements of TWM as trial goals, Table 5-5 details how these indicators might be considered.

TBL results in a region

If a given organization goes the extra mile in its reporting, it can explain impacts on waters within its sphere of influence, but what about its neighbors? If the spheres of influence in a region do not reach to cover all water issues, gaps will occur. Regional TBL reports could address these gaps and account for the myriad small actions as well. To prepare a regional TBL report would address the need for geographic integration in TWM.

Table 5-4. TBL achievements by issue

Issue	Examples of TBL achievements
Economic	Hold rates down
	Improve infrastructure
Environmental	Cut energy use
	Provide instream flows
	Do not exceed wastewater permit
	Improve stormwater runoff with best management practices
Social	Add recreation to stormwater corridors
	Provide outdoor activities around supply reservoir
	Provide outreach to schools

Table 5-5. TWM elements and possible indicators

TWM element*	Possible indicator
Stewardship for the greatest good of society and the environment	An index of organizational and citizen stewardship efforts
Promote supply development	An indicator of firm yield or water reserves
Promote source protection	An assessment of the protection of sources
Enhance water quality and quantity	Integrated report of quality and quantity
Address cost-effectiveness	Rates and investment summary, financial overview
Address environmental benefits and costs	An index of environmental achievements
Promote water conservation	A report of water conservation results
Promote reuse	A reuse ratio
Foster public health and safety	A risk report including health and security
Address social values and foster community goodwill	A report on how well utilities get along and whether water management is cooperative or confrontational, an indicator of corporate social responsibility

*See chapter 3 for an explanation of these elements. Notice that most possible indicators require integrated judgments themselves, and some might be controversial.

The problem is that in most cases no authority exists to prepare regional TBL reports. If a watershed does have a planning authority, such as a river basin commission, then responsibility is clear. Even then, it will probably not be able to address every small watershed, and local watershed groups are needed. Thus, the logical way to handle regional TBL monitoring and reporting is through river basin and watershed authorities and associations.

These do not exist everywhere and, where they do, they are not always able to address issues effectively. When the United States tried to organize river basin commissions across the country under the WRPA, it did not work well.[4] The nation does have today a large number of independent watershed organizations. City and county governments also have a role. Each one could have a watershed unit to assess problems within its own geographic boundary. Soil and water conservation districts can also fill the bill in many cases. Effectiveness of all of these units and their engagement in water quality and hydrologic modification is probably the main question.

TWM requires collective scorecards to track how well it works; otherwise, how can society track sustainability on a broad basis? Scoring TWM for a region is like keeping score of multiple games and teams in different leagues and reviewing the consolidated statistics of all the teams, rather than those of just one team or player. The scorecard tracks how well the game is being played collectively rather than how a single team is doing. Since we naturally root for one team, rather than all of them, this makes it harder to galvanize people's interest. This institutional problem of *regionalism* in any public policy arena is based on our natural instinct to compete rather than cooperate.

In a region, financial and economic results would show on a collective basis how water actions affected the regional economic picture. For example, in the European Union, regional water conditions are to be reported through a competent authority. In another example, a current project to bring water from the Yangtze River to northeast China aims to have dramatic effects on the receiving region's economy. Environmental and social effects on a regional basis are more dramatic. In any case, the entity sponsoring such a high-impact project would have the basic responsibility for TBL reporting about the project.

Regional TBL reporting requires someone to take the lead

The economic and social effects of TWM activities of recognized regional water entities might be apparent, but the environmental impacts of water management might escape the reports of water entities. This is because of aggregated effects from diversions and point discharges of industries, nonpoint sources, and hydrologic modifications or alterations. Accounting for these requires some sort of regional environmental report.

Who is responsible for regional reporting? This is the general dilemma of regionalism. There is an opportunity for utilities to take leadership in organizing regional reporting by contacting organizations with related

---

4 See chapter 10 for an explanation.

missions and setting up programs for coordinated reporting on a regional basis. This reporting should also account for the many smaller and dispersed entities and forces that fly under the radar screen of the utilities and formal reporting authorities.

Compiling a TBL scorecard: the Sydney Water example

Let's look at a TWM scorecard for a utility. A good example is how Sydney Water (2005) presents its TBL results in its annual *Towards Sustainability* report. The report presents "an integrated sustainability reporting format, covering the Corporation's environmental, social, and economic performance."

Sydney Water is a state corporation and is owned by the people of the state of New South Wales. It is governed by a board of directors whose duties are prescribed by the State Owned Corporations Act of 1989.

Sydney Water operates in one of the fastest-growing regions in Australia, an area bounded by beaches, rivers, and Australian bushland. It operates under a license issued under the Sydney Water Act of 1994, and it is regulated by the Independent Pricing and Regulatory Tribunal (IPART). Sydney Water provides water, wastewater, and some stormwater services to four million people. Sydney Water's performance under its license is assessed annually by IPART and reported to the Minister for Utilities and Parliament. The operating license requires performance monitoring and regulatory drivers to require Sydney Water to maximize community investment for objectives including protecting public health and the environment and being a successful business.

Sydney Water's TBL reporting procedures help it to comply with its obligations to its regulators. In the United States, the mostly government-owned utilities do not face such clearly defined requirements, but the sum of their regulatory requirements may amount to about the same.

Sydney Water's statement of multiple objectives is, "We are committed to achieving sustainability in the conduct of our business: operating efficiently to deliver quality water and wastewater services to a growing population, while protecting our unique environment, and acknowledging our responsibilities to the communities we serve" (Sydney Water, 2005).

Its strategic plan, WaterPlan 21, calls for a sustainable and integrated approach with a wide range of strategies, programs, and projects. These are to consider:

- Integrated management of water, wastewater, and stormwater;
- Efficient use of water and reduction of demand, rather than building new dams;
- Use of sewage as a resource and not a waste;

- Being environmentally sensitive and economically efficient; and
- Roles of individuals in achieving sustainable water management.

Results sought are: clean, safe drinking water; sustainable water supplies; clean beaches, ocean, rivers, and harbors; wise resource use; and water-smart growth.

Sydney Water's strategic plan, along with its TBL reporting system, provides an integrated approach to planning and accounting for sustainability. Of course, the regulatory oversight that is built into their license requirements seems to help create incentives for their comprehensive approach.

A US example: Seattle Public Utilities

A large US utility on the scale of Sydney Water would have to decide itself to do TBL planning and reporting. All of the elements are in place for them to do so, and some utilities have mounted impressive programs. The approach taken by Seattle Public Utilities (SPU), which has similarities to Sydney Water but serves a smaller population (about 1.3 million receiving some water services), is an instructive example (2007). Its approach could serve as a model for other US utilities.

SPU has a Strategic Business Plan and an Environment Report. Naturally, SPU recognizes that its first responsibility is to its customers but that it has environmental and social responsibilities as well. It recognizes this in its strategic plan, which has four consistent key statements. To see how these fit with TBL planning, Table 5-6 displays the statements with comments about the TBL and TWM.

These principles of SPU's operation are embodied in its mission statement: "We provide our customers with reliable water, sewer, drainage and solid waste services. We protect public health and balance our social and environmental responsibilities to the citizens and community, while providing cost effective service to our ratepayers."

Sydney Water's TBL report and Strategic Plan are linked

In its environment report, SPU explains how it meets environmental challenges and commits to take on future environmental challenges; manages water resources with a water conservation program, watershed management plans, a unique habitat conservation program, and new approaches to improving water quality; leads the region with drainage and wastewater initiatives to protect urban creeks; protects salmon and other wildlife; and incorporates its vision into its everyday business practices. The report is dated 2001, which indicates that SPU does not believe it must issue the environmental report every year.

Table 5-6. Seattle Public Utilities statements and TBL/TWM

SPU strategic statements	TBL/TWM comments
Our most important responsibility is providing basic utility service	A utility must concentrate on this basic mission and work on environmental and social outcomes in parallel with it
Customer service is a key to our success	A strong customer focus is essential to water service organizations
Employees must be provided with resources to do their jobs	This is embodied in the TWM principle to enable the workforce
We are changing the way we make all decisions by incorporating social, environmental, and financial outcomes and benefits	This is the key TBL/TWM strategy, to focus on social and environmental goals as well as, but not instead of, business objectives

Source: SPU, 2007.

Integrity in reporting

One of the issues to face in TBL reporting is how to maintain integrity in the process. It is easy to take your eye off of the ball when you are dealing with intangible issues, as compared to bottom-line financial issues.

In financial reporting, integrity and the quality of information are controlled by rules and regulatory agencies, such as the Internal Revenue Service, the Financial Accounting Standards Board (FASB), and the Securities and Exchange Commission. No such regulatory apparatus exists for TBL reporting. Actually, water agency reporting is mostly unregulated, except for legal requirements and political and watchdog processes.

> **Accountability and integrity in TBL reporting are essential**

There is no easy answer for how to ensure integrity in the TBL reporting process. At the end of the day, the oversight body of the water service organization, working with the public and the press, must take responsibility. A TBL report should not be primarily a public relations effort, but it is undeniable that issuing one will have effects on the public image and hopefully on public stewardship.

Summary points

- The broad goals of TWM and its focus on sustainability suggest that Triple Bottom Line (TBL) reporting can be used to keep score.
- TBL reporting is a form of multicriteria decision analysis (MCDA) and includes information on economic, environmental, and social accounts of a water utility. The economic accounts address financial results and economic development. Environmental and social accounts address

positives and negatives for habitat, society, and related issues. TBL reporting enables utilities to track how they are doing in corporate social responsibility.
- TBL reporting originated among sustainable development groups and is a way to improve corporate governance and transparency in the economic, social, and environmental arenas.
- The tool of multicriteria decision analysis provides a framework to organize evaluation information for these categories of goals and to study their trade-offs.
- The science in TBL reporting is in compiling the numbers and displays, and the art is in deciding what is important and how to present the information to have the best effect.
- While it has effective financial indicators, the water industry lacks consensus indicators for social and environmental accounts that can be used in TBL reporting. These must be developed on a case-by-case basis.
- TBL reporting can lead to more trust among staff, customers, regulators, and stakeholders on sustainability issues; help in attracting and retaining employees; and lower risk profile through the sharing of information.
- The water utility must work with oversight bodies, the public, and the press to ensure the integrity of its TBL reports, raise its public image, and strengthen its corporate stewardship.

Review questions

1. How does TBL reporting differ from current approaches to reporting of results by water utilities?

2. Explain how TBL reporting relates to the practice of multiobjective planning as it has evolved in the United States.

3. Is TBL reporting related to the concept of corporate social responsibility? If so, how?

4. How can the tool of MCDA be used to aid in TBL reporting?

5. What might some benefits of TBL reporting be for water utilities?

References

Alegre, H., J. Baptista, E. Cabrera Jr., F. Cubillo, P. Duarte, W. Hirner, W. Merkel, and R. Parena. 2006. *Performance Indicators for Water Supply Services.* 2nd ed. London: IWA Publishing.

Battle for Brainpower. 2006. Special insert, *The Economist* 381(8498): 3–24.

CERES. 2006. *About Us.* http://www.ceres.org/. Accessed October 5, 2007.

Global Reporting Initiative. 2006. *Who We Are.* http://www.globalreporting.org. Accessed October 6, 2006.

Grigg, Neil S., and E. Vlachos. 2005. *Condition and Security Indicators for Interdependent Infrastructure Systems.* Final report for the National Science Foundation. Fort Collins, Colo.: Colorado State University.

Kaplan, Robert S., and David P. Norton. 1996. *The Balanced Scorecard: Translating Strategy Into Action.* Boston, Mass.: Harvard Business School Press.

Kenway, Steven, and Linda Reekie. 2006. Triple Bottom Line Reporting: A Concept to Support Water Utility Sustainability. AwwaRF. *Drinking Water Research* 16 no. 1 (January/February): 5–8.

Niven, Paul R. 2006. *Balanced Scorecard Step by Step: Maximizing Performance and Maintaining Results.* New York: John Wiley & Sons.

Seattle Public Utilities. 2007. *About Seattle Public Utilities.* http://www.seattle.gov/util/About_SPU/index.asp. Accessed May 28, 2007.

Stahre, Peter, Jan Adamsson, and Örjan Eriksson. 2000. *Performance Indicators for the Water Industry: An Introduction.* VA–FORSK Report 2000.8. Stockholm: AB Svenska Byggtjäns.

Sydney Water. 2005. *Sydney Water Annual Report 2005: Environmental, Social, and Economic Performance.* http://www.sydneywater.com.au/publications. Accessed October 6, 2006.

Tampa Bay Water and CH2M Hill. 2006. *Decision Process and Trade-Off Analysis Model for Supply Rotation and Planning.* Denver, Colo.: AwwaRF and AWWA.

US Environmental Protection Agency Watershed Academy. 2006. *Introduction to the Clean Water Act.* http://www.epa.gov/watertrain/cwa/cwa25.htm. Accessed November 17, 2006.

CHAPTER 6

VALUE AND COST OF WATER

In the water business, it is pretty well-known that if something is almost free, people will waste it. If something is perceived to have value, people will be better stewards. Water has a special problem because many people believe it should be free. So how can we overcome that dilemma and get everyone to value water?

The issue is captured by a story told by Tracy Mehan (2007), former US Environmental Protection Agency (USEPA) assistant administrator for water. A nun protested a rate increase by saying that since God provided the water, it ought to be free. The water manager replied, "Sister, we agree the water should be free, but who will pay for the pipes and pumps?"

Valuing water—a core issue of TWM

Who will pay for pipes and pumps is one issue. The other is that if people waste water, less is available for other uses, especially environmental uses. If people do not appreciate the value of water, they're more likely to waste it. So valuing water has both a financial aspect and the economic aspect of how to conserve and allocate it.

> **Yes, water is free, but who will pay for the pipes and pumps?**

If people waste water, they deprive higher-value uses. However, this issue goes much further. If people think water is plentiful, they may not understand the need to conserve and pay for it so that there are supplies for others to use. For example, people with houses on a lake may expect water levels to stay high so they can enjoy the view but not appreciate that the water level must fall so water can be sent to downstream users. They perceive only their loss and not the gain to others.

The American Water Works Association (AWWA; 2005) has picked up the importance of valuing water. Its Strategic Plan states that it will "engage the public, elected officials, and key decision makers about the *value of water* and AWWA's role in maintaining that value" (emphasis added). This key phrase has been included because AWWA sees that valuing water is the key to efficiency in its use.

TWM also focuses on the value of water through its aim to balance "competing uses of water through efficient allocation that addresses social values, cost-effectiveness, and environmental benefits and costs." The efficient allocation must be based on some system of relative values so that decisions can be made. How to value water is a core issue of TWM if it is to balance among social, economic, and environmental uses.

This chapter focuses on economic valuation, and chapters 7 and 8 explain environmental and social values. The three come together in Triple Bottom Line planning and reporting, where the balance is achieved.[1]

A fictitious water market

This problem can be illustrated by the fictitious Water Market and Auction shown in Figure 6-1. Imagine that we have a pure market system for allocating water, and that all the uses are represented at the market auction. The market is running a regional water system that consists of a river and a lake. Each use is represented by a bidder at the auction who brings to the sale enough money to purchase the water for that use.

A game needs a scorecard

The water utility takes water from the lake and must buy enough for the city supply. However, the water quality manager must make sure enough is left in the stream below to dilute the wastewater. The water quality manager, the trout guides, and the kayakers soon realize they are bidding on the same thing—water to be delivered down the river—so they get together and form an *instream flow bidders coalition*. They are approached by the electric utility and the industries, who both tell them they also deal with released water and that they can probably get together and work something out. Soon, these instream flow interests have worked out a coordination scheme whereby they can purchase a block of water to be released and they can have an internal system to allocate it among themselves.

---

1 Triple Bottom Line planning is discussed in chapter 5.

Figure 6-1. A fictitious water market and auction

An argument breaks out among the flood insurers and the lake recreation company because one wants to keep the lake water low and the other wants to keep it high. They soon realize that not only is water for sale but also storage capacity in the lake. While they are bartering over this issue, the instream flow coalition and the water utility join in and inform them that they require guarantees from the lake as well because they don't want to be damaged by drought.

Pretty soon, all the bidders realize this is more complicated than they thought, and they set up a formal market to work things out. It resembles the futures market for agriculture and commodities so that water and reservoir space can be sold now or in the future. Shares can be turned in anytime for actual delivery of water according to the schedule in the purchase. All of the water users are happy with the system, but soon some of them realize they are easily outbid by the others. The electric power utility can easily buy as much water as they want, but the trout fishing guides and the kayakers can hardly afford to buy any water. Another problem is that they can't agree at all on the schedule for water flows.

A fully functioning water market would be a complicated thing

The trout guides lodge a protest and tell the market managers that leaving water for trout ought not to follow the same rules as the other uses, because if water is left for trout, it also nourishes beaver and geese, which depend on the river as well. People like to look at the beaver and geese, so you have extra value from trout water. The market managers say, "So what—who will pay for that?"

The market continues to operate, but soon the many conflicts and complexities of the real world of water management are at play, and the bidders realize that their market system is not adequate by itself to handle all the value problems. So they decide to set up a nonmarket side to the system, and they need a governing board for that. Soon there are big arguments over how to select the members of the board to represent the many different interests at play. They finally get a board appointed, and it helps out with some of the problems, but it creates a lot of new political problems that were not apparent at the beginning.

The challenge of valuing water

This story shows how water has many values in its uses for drinking, irrigation, recreation, industrial production, and other uses, but people don't always see these values the same way. In theory, the best allocation of the water returns the highest value to society, and this requires that values be assigned to the different uses. In the end, decisions about water allocation are more political than they are economic, and if water management is to work well, the public must realize the value of water and be willing to pay for it. This is not to say that we do not continue to try to price water's use. For example, the Water Framework Directive of the European Union requires countries to ensure that water pricing policies provide incentives to use water efficiently and for sectors to contribute to full recovery of costs of water services, including environmental costs.[2] Other nations also continue to consider water pricing. For example, Colombia has an ambitious program to recover costs of both water abstraction and discharge of wastewater.

How well we succeed in engaging the public, elected officials, and key decision makers about the value of water depends on our understanding of what this means. This chapter explains what it means so that the dialogue can be informed. It explains how to determine the need for water in various uses so that values and costs can be assigned to provide a balance.

The chapter also probes why people do not recognize the value of water and what can be done to address this problem. It explains how to assign value to the allocation of water to water supply, irrigation, wastewater disposal, and instream uses such as hydropower, recreation, and navigation. Valuing of flood control is different because we are valuing the prevention of damages, but the beneficial use of water to clear out floodways and nourish wetlands is a positive flood-related water use benefit for environmental uses.

---

2 See chapter 3.

How society balances the allocation of water resources

In a perfect world, society would balance the allocation of water by applying economic theory, where allocation occurs according to how much benefits exceed costs. In practice, the process is not that way—things are worked out with politics, money, and lawsuits. One reason for this is that most things are done by politics and law, not by economics. Another reason is that we simply lack the economic tools. This is explained by Robert Young (1996), a water economist who wrote a guidebook on water pricing for the World Bank. He explained how market prices for water are seldom available and how water-based environmental effects are seldom priced.

In graduate classes, we learn the theories of economics as they apply to resource allocation, and we study interest rates and benefit–cost analysis. However, we seldom see a decision in the real world that works like that. Soon most students lose interest, and they say that economics seems abstract to them.

The details of valuing water as a resource can be abstract, and some of it is not very useful other than in a theoretical way. What is important is to find practical uses for the economic concepts of value to help society balance allocation of its water resources. It works at two levels to do this, and the levels depend on the resources that a particular decision maker controls.

At one level, an individual utility controls its own assets and the rates it charges its customers, so it allocates its resources to maximize return on assets as it perceives its responsibilities to its customers and stakeholders. At another level, a higher authority must allocate water among utilities and organizations, and it requires a uniform standard of value so it can decide. These are the two basic ways that society values its water resources.

At this utility level, a decision might be needed as to how to set rates between residential and commercial users, with the goals to conserve water and to bring in enough revenue to run the utility, all the while being fair to all parties. In setting its rates, the utility might have to pay environmental mitigation costs so that it is, in a tactical way, valuing the use of water as an environmental resource as well as the use of water for its customers.

> **Unfortunately, water decisions seldom follow classroom theory**

The higher level of allocating water among different kinds of uses was explained by Raucher et al. (2005), who showed how assigning value to water uses helps us identify strategies that provide for the greatest amount of total well-being for all members of society, or maximizing social welfare.

They are addressing the level at which society is allocating among water users, as opposed to the lower level where a utility is maximizing its own interests. This higher level might seem somewhat theoretical compared to the level where a utility actually makes its decisions, but the two levels are really interconnected. When a regulatory agency issues a permit or makes a similar decision, they are implicitly valuing the outputs from water to the extent they can. They would like to have well-defined rules, of course, because that takes the onus off of them and they can shift any conflict to some other authority. Rules can be that authority.

The definition of TWM emphasizes this higher level in seeking the balance in use of water. Actually, how society balances use of all of its resources based on value is a central issue in economics and has been studied by more than one Nobel Prize winner. Given its importance, water management has provided research topics in economics that have led to a number of advances, including benefit–cost analysis (BCA) itself. Early work dates back to the 1930s and continues to provide rich material for today's public-sector economists.

To get the balance right requires options for allocating resources, ways to value these options, and a decision-making mechanism. Options for allocating the resources are proposed by utilities and organizations based on their own assessments of how they meet their obligations in cost-effective ways. The selection among these is then based on politics, which is based on what the community values as a whole and reflects on the basis of its votes.

Figure 6-2 shows how, at the level of utilities and organizations, each entity decides on its best course of action for water uses and related valuation decisions. At the higher levels, an authority decides on the allocation of resources. At these levels, BCA might be used to identify how the best uses of society's resources can be achieved. At lower levels, cost-effectiveness is used as each organization seeks to optimize its use of the resources it controls.

Figure 6-2. How valuation of water differs by level

We can summarize the process by saying that each decision-making level allocates the resources that it controls to achieve its own purposes as it sees them. Society wants to look at things from the standpoint of total human welfare, and organizations want to maximize return from their own assets. As an example, say multiple water users in a basin have permits for different uses. A drought occurs, and regulators must allocate the scarce water over the uses. On what basis does the regulator have to decide? Some valuing system or system of priorities must be used. This is a higher-level decision. The lower-level decision would occur for example when an individual utility decides to ration its available water among uses with some charging scheme.

Opportunity cost is an important tool to help decide on the allocation of resources that you control. It means the cost incurred or benefit lost by giving up one option so that you can exercise another one. For example, if you own a farm and you convert it into a residential subdivision, you lose the opportunity to farm.

In the case of water, its use in one application usually takes away from another one.[3] For example, if you divert water from a stream to use it for urban water supply, you lose the opportunity to enhance an instream fishery. As another example, a transbasin diversion of water permanently deprives the basin of origin of the right to use and develop the water. So to compute a net benefit from some water decision, you would take the total apparent benefit such as revenues from water sales and deduct the benefit lost when you give up the opportunity to apply the water to a fishery.

In water management, each authority allocates the resources it controls

Although they are not perfect, tools are available to assess whether a water decision meets the TWM requirement that it is "for the greatest good of society and the environment." The tools include economic impact evaluation and benefit–cost analysis, environmental assessment and impact statements, and social impact assessment.

Economics and finance—deciding and paying

An important goal of this chapter is to make a complex subject clear so that it can be applied by managers. Toward that end, it is important to distinguish between water economics and water finance.

Economics is used for analysis and decision-making, whereas finance is a focused management tool. Both fields use monetary values, but finance

---

3 Economists call these "rival" uses, as opposed to "nonrival" uses (see Young, 1996).

deals with how to pay for things and economics deals with decisions about allocating resources (Figure 6-3).

Economics helps society find a balance in the use of its resources. It uses tools such as benefit–cost analysis and environmental impact statements. *Finance* involves practical decisions based on the bottom line—profit or loss, rate of return, and so on. It uses budgets, revenue and cost analysis, and financial statements.

Value and cost-effectiveness are economic concepts aimed at identifying the right decisions by society. Individual utilities make their decisions on the basis of finance, rather than what is good for society as a whole. Here we have a practical aspect of the "it's not my problem" syndrome. Concepts such as sustainability and TWM are for the good of society, but the primary incentives of utilities focus on organizational self-interest.

The distinction between economics and finance in water management is therefore important and fundamental. Economics is used in coordination mechanisms to achieve balance under TWM and water planning. It produces information on the value of water that can go into TBL reports. Finance is used to figure out how to pay for water and infrastructure, and its results are reported in accounting statements. Naturally, a utility will emphasize its financial management to ensure its continuing operations, but financial management is not the core issue of TWM.

What is meant by the value of water

The value of water is actually a fundamental and encompassing issue of how society decides who gets to use water and how it is used. The practice of Total Water Management, as stated in its definition, is intended "for the greatest good of society and the environment." This could also say that TWM should lead to the greatest *value* to society and the environment. By replacing *good* with *value* we make the definition more specific.

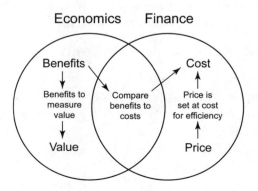

Figure 6-3. How economics and finance differ

General concepts of the value of water

Value is one of several related monetary concepts that explain the economic attributes of water. Others are cost, price, and benefit.

Value is the worth of something. What has value to someone depends on what is important to them. What is important depends on their underlying priorities or "values."[4] People have different priorities, beliefs, cultures, and underlying values, so they do not agree on how to allocate society's resources. This is why the political process is necessary as a mechanism to balance their interests.

The factors that determine someone's underlying priorities can include a range of motivational variables from desire for survival to higher-level concepts such as natural beauty. All of these are used in water resources decisions that consider economic, social, and environmental impacts of water uses. In some cases, they can be quantified. For example, risk of death from a flood can be estimated, if not with great precision. Intangible values such as natural beauty are much harder to assess.

> Economic concepts related to water include value, cost, price, and benefit

Economists measure value by willingness to pay. However, someone may value something but think someone else should pay. For this reason willingness to pay can be a narrow concept that applies mostly to private goods. Public goods are often paid by society as a whole through taxes, and willingness to pay is not as relevant as political choice in making decisions.

The concept of *benefits* has been introduced to aid in political choice about resource allocation. As used in economic evaluation, benefits include economic, social, and environmental categories. *Cost* is the expense to produce something and *price* is the amount charged for the exchange of a good or service. *Value* is a measure of a person's perception of the overall worth of something.

So the four concepts are related but distinct. As Figure 6-4 illustrates, benefit is used in economic evaluation, usually to measure value perceived by society, rather than to a single consumer. Value as a measure of a consumer's willingness to pay focuses more on the market situation in which a person's willingness to pay initiates a transaction. Cost is mostly a financial concept but also is used to compare societal benefits and costs. Price is also mostly a financial concept but also is used in economics to signal the setting of prices at the level of costs to achieve economic efficiency.

---

4 This is another meaning of the word *value*.

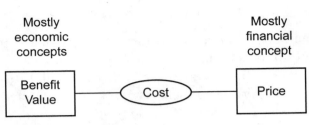

Figure 6-4.   Concepts of value

Societal versus individual decisions: the accounting stance

How economic, environmental, and social values are estimated by society and by individual organizations are macro- and micro-level issues. The "it's not my problem" syndrome results from the gap between public interests and the narrower interests of organizations and private individuals. So it is useful to look at decisions from the perspective of the *accounting stance,* meaning how the entity views the benefits and costs. Three such accounting stances come into play: society, the organization, and the individual.

The organization or an individual will compute its own benefits and costs and will focus on its financial bottom line. Water decisions by organizations and individuals affect society as a whole, and TWM addresses them by shared decision-making.[5] Utilities will focus on their bottom lines, the reduction of risk, and sustaining their operations. They will consider local environmental issues, especially related to resources they control or are responsible for. In the social realm, utilities will be aware of local public health, customer service, people with special needs, and community issues.

At the macro level, society looks at the good for all and the rhetoric is different. Table 6-1 illustrates how society and organizations view these different categories.

The difference in society or organizational views of benefits and costs is due to their different accounting stances. Society's accounting stance is where benefits and costs are compared according to society as a whole. This is the issue with benefit–cost analysis, which was introduced through the Flood Control Act of 1936. This act required the comparison of benefits and costs regardless of who they accrued to; it is a society-level view. Utilities and other organizations compute benefits and costs from their own accounting stances, meaning how they come out ahead or behind.

---

5   See chapter 4.

Table 6-1. Issues as perceived by society and organizations

Issue	Society	Organizations
Economic	Macroeconomic, federal tax revenues, government finance	Microeconomic, local taxes and fees, organization finance
Environmental	Large-scale effects, enforcement of environmental laws	Local environment, citizen attitudes, compliance with regulations
Social	Social well-being of nation, political goals of equity	Meeting citizen needs locally

In national economic accounting, analysts can be more theoretical about economic estimates than they can with local issues. Economists and government officials pay close attention to correct economic analysis methods, but in the final analysis, local decisions in the political process may or may not pay any attention to these higher-level views.

Utilities and local organizations cannot afford to be theoretical about their estimates because their ratepayers and taxpayers want to know how much it will cost and who will pay the bill. Bringing decisions close to the local level in this way is good in that it provides a measure of fiscal discipline. The challenge is to take care of the broader needs that fall outside the narrow interests of the local jurisdictions.

A single organization or even an individual can perform a benefit–cost analysis, but for what purpose? It depends on the mission. For example, a water supply utility could estimate the benefits from developing a new water supply, but it normally will set a target for supply and look for the least-cost method to find it. It normally will not compute a benefit–cost ratio (BCR).

The difference between societal and individual decisions explains much of the need to provide a balancing. In addressing how society and its decision makers should choose among alternative ways to manage water, TWM aims at the societal level. However, it must also work for organizations. How diverse stakeholders place value on outcomes of decisions is important because they value societal outcomes differently. This is where balancing and conflict management are needed (Figure 6-5) through the TWM process.

The accounting stance is an important determinant of water resources decisions. At the national level, economic estimates look at the benefits and costs regardless of who they accrue to. At the local level, utilities must explain to their own ratepayers and taxpayers what they are getting for the money they pay.

Figure 6-5. Water management is a balancing act

How society computes benefits and costs

The theory of benefit and cost accounting is that for any proposed action the positive impacts (benefits) and negative impacts (costs) are estimated and compared. There must be standards or guidelines for how these impacts are estimated, or there can be no basis for comparison. This was the purpose of the Principles and Standards (P&S) that were developed under the Water Resources Planning Act of 1965.

In the P&S, three economic concepts were recognized: national economic development, regional development, and distribution of benefits and costs. National economic development measures the size of the national pie or economic efficiency. Regional development measures the same thing, but for a region. Distribution of benefits and costs measures economic equity or how the pie is sliced up.

How to assign value to uses of water: computation of benefits

Some 70 years ago, water planners and economists developed the tool of benefit–cost analysis (BCA) as a way to compare the payoffs of projects to their costs. Actually, water planners get the lion's share of credit for developing BCA, which has now been extended to analysis of other types of government programs.

In the BCA for a water project, benefits and costs are measured by dollars for each category of water use in the project. Once they are reduced to the same time basis and made commensurate, benefits and costs are compared on the basis of the benefit–cost ratio, net benefits, or their rate of return.

> **The tool of BCA was developed in the first place for water decisions**

While decision makers do not always accept the results of BCA because of difficulties in estimation of benefits and costs, if consistent techniques are used, projects with greater merit show up better on a relative basis.

Benefit–cost analysis is sensitive to the interest rates used to make monetary sums commensurate with each other. Whereas in financial analysis the interest rate is the cost of money, economic analysis requires that social purposes also be considered and is more difficult if not impossible to determine.

Value of water in different uses

Benefits are computed differently for the different uses of water. Examples are municipal and industrial water supply, irrigation, hydropower, navigation, and recreation. Flood damage reduction from a water project is also a benefit of water management, but it focuses on prevention of damage rather than on supplying a commodity.

A number of documents provide guidance on estimating values or benefits from water uses. Howe (1971) is a good general reference, while the Principles and Standards (Water Resources Council, 1983) comprise a textbook on estimation of benefits. The AwwaRF report by Raucher, et al. (2005) and entitled *The Value of Water: Concepts, Estimates, and Applications for Water Managers* provides a good overview of recent estimates of water's value. Young (1996) explains further details of nonmarket methods of estimating water's value.

This section reviews methods for benefit estimation for most uses of water. To begin we need a clear picture of what constitutes a *benefit*. Basically, a benefit from an action or project is a new and positive contribution toward a goal. It should result from the action, and without the action, it would not occur (this requires a *with and without* analysis). On the other side of the ledger, the *cost* is a negative result from the action.

In general, there are two ways to estimate benefits. One is to estimate the market value of a product that results from a use of water, and the other is to estimate as a benefit the cost of the next most expensive alternative way to provide the water needed. Estimating market value is intuitive, but the other approach is less so.

Estimating as a benefit the cost of the next most expensive alternative way to provide the water needed for a particular purpose provides a convenient way to compute net benefits, but the information provided may not be very useful. Say you need to construct a water project to serve a particular need, and it can cost a sum of $1 million, $2 million, or $3 million. The least expensive project costs $1 million, and the next least expensive $2 million. If you choose the least expensive project, and the benefit is the next least expensive cost, or $2 million, then the net benefit is $1 million and the BCR is 2. This almost seems like an artificial way to estimate benefits and amounts to little more than to say you have decided to do something and will do it as cheaply as possible. Thus the benefit (or savings) is measured by the next least expensive option.

Both methods (market value and next least expensive option) are complex and difficult, but a few examples should clear the air on them.

Municipal and industrial water

The positive you get from a water supply project is the new water, which has value. So one way to value the water would be the amount you can sell it for. Thus, if a project had an annual yield of 1,000 acre-feet of water and you could sell it wholesale for $200 per acre-foot, the benefit could be estimated at $200,000 per year.

However, because water supply is sold at rates charged by a utility, and because a utility sells the water for its *cost of service,* this approach leads to the conclusion that benefits equal costs because we made them equal. Water rates charged by utilities can be set at cost of service because the utility has a monopoly. When a utility is regulated by a regulatory commission, this cost of service comes under the scrutiny of cost accountants. However, most urban water service is supplied by local government utilities, and the only rate scrutiny received is through customer reactions and the political process.

Historically, residential water rates were rather low, reflecting the utility's cost of service to provide the water. Although AWWA surveys the charging systems used by utilities, it is difficult to compute an average customer cost per thousand gallons. As an example of today's rates, for single family residential taps my city charges a monthly conservation rate comprising $1.87 for the first 7,000 gallons, $2.15 for the next 6,000 gallons, and $2.48 for all over 13,000 gallons. If a family of three used 150 gallons per capita per day (gpcd), then for a 30-day month their total charge for water would be $27.23. This works out to be an average cost of $2.02 per thousand gallons. This is probably a representative cost from around the nation.

The true value of water in residential use is not what a utility charges to meet its cost of service. A better measure of value of residential water might be what people will really pay for it. People seem willing to pay more for residential water, especially for drinking. According to Raucher et al. (2005), evidence is mixed, but the public is willing to pay prices for water that might reach $4,000 per acre-foot ($12.28 per thousand gallons) or higher[6]. If a 16-ounce bottle of water costs $1, the cost is equivalent to $8 per gallon, far more than gasoline

---

6 $1 per thousand gallons = $325.85 per acre-foot. $1000/AF = $3.07/TG.

costs. These are much higher than today's typical utility charge of $2 to $3 per 1,000 gallons.

The other method to value water by estimating benefits is to recognize that water supply is generally provided because there is a demand for it. The analysis of its benefits can be based on a requirements approach (James and Lee, 1971). In this approach, the benefit is computed as the cost of the next least expensive alternative to meeting the demand. The theory here is that because you have decided the project is needed, a valid benefit–cost ratio can be computed by dividing this next-least-cost project by the least-cost project (Howe, 1971).

This means that if a water project costs $2 per thousand gallons (TG) to produce water and the next least expensive costs $2.50/TG, the benefit is $2.50/TG and the cost is $2. Thus, the net benefit is $0.50/TG and the BCR would be $2.50/2 = 1.25. This approach adds no real information other than how much financial advantage the least-cost project brings over its nearest competitor. It also depends on the predetermined decision to develop the water, one way or another.

Water supply rates reflect a utility's cost of service because utilities are monopoly enterprises that must cover their costs. It can be argued that their cost of service is low because they do not pay the full environmental cost of the water. If the utility is required to meet society's opportunity costs of using the water in other ways, then the rates will reflect the full cost of water on a sustainable basis.

Agricultural water

Applying water for irrigation seems a more logical way to use increased production from water as a measure of the benefit. Let's say that you are a corn farmer and can grow 100 bushels per acre just from rainfall, but if you irrigate, the yield will be 150 bushels per acre. The benefit will be the increased income from the greater yield.

If the calculations were that simple, we could compute water's value for irrigation directly. However, it is more difficult than that to assess the value of water applied for irrigation. In the West, it is common for people to say something like, "The West has a water shortage, and 90 percent of the water is applied to low-value crops such as hay and corn." They might add, "All the West has to do to meet its water needs is move water from agriculture to cities."

To consider the merit of this statement, we should ask, "Why is it that a water-short region like the West has built all of these reservoirs and irrigation structures to grow low-value crops?" The answer is complex, but one part of it is that it was national policy to subsidize irrigation in the West to stabilize settlement there, and that's a main reason the

Bureau of Reclamation was organized in the first place. The involvement of government in irrigation still distorts values and makes it difficult to apply simple market analysis to estimate the value of crop water.

Setting aside any conclusion about the merit of the statements about irrigation water being wasted, it is easy to compute whether the economic production from a crop is less than from water used in cities. Say irrigation increases the yield of an acre of corn from 75 to 150 bushels. At $4/bushel the irrigation adds $300 in revenue per year. Now, if it takes 3 acre-feet to irrigate that acre, the annual benefit from each acre-foot is $100. If that same acre-foot was used in urban water supply at $2 per thousand gallons, it could be sold for about $650.[7] So the initial impression is that the benefit from water in the city is six times that of corn.

> **Why did a water-short region like the West build reservoirs and irrigation systems to grow low-value crops?**

It is, of course, more complex than that. Farmers might say, "We have been cut off from water, and the people in the city can still water their lawns." This implies that the farmers think their crops are worth more than the lawns in the city because the crops are their livelihood, whereas urban lawns are an amenity.

In the past, big government irrigation schemes in the West offered concessionary terms and subsidies because there were multiple reasons to develop irrigation, including stability and settlement of the region. These projects have already been built, and no new ones are likely to be approved. The future for irrigation water may be toward smaller schemes, where more straightforward analysis can be applied.

If, for example, a greenhouse operator installed an irrigation system, the analysis would be the same as for any capital expenditure by a business. If a developer uses irrigation for golf course maintenance or for landscaping projects, the benefits would normally be significant.

Wastewater and water quality

When water is used to dilute wastewater and improve stream water quality, the benefits are hard to compute because they deal with improvements to intangible goals such as public health, environmental habitat, and aesthetics. The nation has decided through legislation that these are worth the cost and does not really apply benefit–cost analysis to decide what to do. Instead, this use of water responds to regulations and command-and-

---

7 This simple analysis ignores the cost side of the irrigation and the urban water system.

control decision-making, not economic decisions. Operators do not have the flexibility to meet standards at different levels, although the concept of trading water quality rights is a possibility for the future.

Hydropower

Analysis of hydroelectricity benefits is similar to that of water supply. You can either apply market prices or consider the next least costly alternative as a measure of benefit. The problems here are the same as they are with water supply, however, and normally a power producer has decided to build a project and will perform a cost-of-service study.

Recreation and fisheries

It is possible but difficult to measure recreation and fishery benefits from water use, and they are hard to justify purely on an economic basis. Say that a lake project is built for $50 million and, considering bond payments and operations and maintenance (O&M), the annual cost is $4 million per year.[8] If that lake served a regional population of 500,000 and, in any given year, 10 percent of the people used it for an average of 5 days per year each, then the use would be 250,000 user-days per year. As you see, each user-day costs $16 and for a family of four would cost $64 per day. That would be expensive recreation.

As another example, say water has to be reallocated from hydropower production to preserve a fishery downstream. The required flow is 100 cubic feet per second (cfs), which under this assumption, is lost to hydropower production. That flow, if saved for use during peak periods, would have produced some $250,000 in power per year, sold at wholesale prices. Say the fish preserved from that sacrifice are a threatened species that is not sought for commercial or sport fishing. If that 100 cfs produces 100,000 8-ounce fish that do not go for any visible economic purpose, the joke will soon be that we are paying $5 per pound for these worthless fish. Obviously, there is something missing in such a naïve analysis.

Navigation

Navigation is another case where benefits are hard to prove or justify. Water-based navigation supports commerce and recreational boating on a system of navigable streams and ports and harbors. For navigation to be reliable, it is necessary to maintain adequate depths in ports and river systems. Waterways and ports require maintenance dredging to maintain

---

8 This hypothetical situation is based on an interest rate of 7 percent, project life of 50 years, and $500,000 annual operation and maintenance cost.

adequate depths and to eliminate hazards to navigation. Locks are required to raise shipping above steep reaches of streams and to move it up- and downstream. Both dredging and the operation of locks have effects on instream flow conditions.

If the boats and vessels plying the streams were required to pay the full cost of developing and maintaining them, the economic cost would be high. This is another area of water management where the government has been active, especially through the US Army Corps of Engineers, which has responsibility for waterways. Recreational boating is often allowed on streams and as an "extra" on lakes that have other primary purposes.

Computation of benefits from navigation is similar to that on other transportation routes. If it can be shown that savings occur by use of navigation, then those savings become a benefit. As an example, if it costs $10 a ton to transport coal by rail but only $5 a ton to transport it by water, then the benefit of using the water is $5 per ton.

Environmental

Environmental benefits other than the economic purpose of fishing are especially difficult to estimate. Chapter 7 explains these, and they include aesthetics, nourishment of all kinds of habitat, vegetation, and even benefits such as the mitigation of the effects of global warming. Although academics do analyze them, there really are no direct and reliable methods for placing values on these benefits. At the end of the day, their value is determined through political and legal processes.

Environmental benefits are important but hard to value

Flood control

Benefits of flood control comprise reduction of economic losses from damage but also the social benefits of reducing the misery of people exposed to flooding. If the people who are flooded have low incomes, the economic damages may seem low, but the misery index may be high. If they have high incomes, chances are the economic damages will be high, but they will not be bothered too much by the flooding because they can afford to escape it. Therefore, the regular economic analysis of flood loss reduction must be supplemented by social equity analysis.[9]

---

9 See chapter 8 for an explanation of social impact analysis.

Use of cost-effectiveness analysis

The definition of TWM says that "Total Water Management . . . balances competing uses of water through efficient allocation that addresses . . . cost-effectiveness." It is well to explain what is meant by *cost-effectiveness*. Actually, the phrase has a precise meaning in economics, which is that once you have decided to do something, cost-effectiveness measures the most efficient and lowest-cost way in which to do it.

Balancing the uses

All water values are to be balanced by TWM as it "balances competing uses of water through efficient allocation that addresses social values, cost-effectiveness, and environmental benefits and costs." This feature means that water decisions under TWM allocate water effectively so that costs are commensurate with all of the benefits and values. The value of water addresses the benefit side of the equation. The other side is how the costs are managed. The discussion up to this point suggests that our system of valuing is not good enough to do this in a completely rational way. Perhaps a simple example will further illustrate this.

Say a reservoir is to be built and it has two purposes: urban water supply and fish and wildlife propagation in the river below the dam. The study shows that to provide the water supply the annual cost is $160 per acre-foot in the reservoir, or about $0.50 per thousand gallons of water stored and delivered. Everyone is happy with that cost. Fish and wildlife costs of the project are about the same, some $160 per acre-foot of reservoir water, and these costs must be provided for in addition to the direct water supply. In other words, the reservoir must be larger so that it is not drained each year to provide the water supply. Someone calculates that only about ten fish per acre-foot will survive in the reservoir, therefore on the average the cost is $16 per fish per year. This catches the eye of an antitax crusader, who complains, thus setting off a political battle over the fish storage in the reservoir. The pricing system breaks down, but the decision still must be made to balance the economic, environmental, and social values.

Why people do not recognize the value of water and what can be done

As we see, the reason that the concept of value of water is easy to enunciate but hard to implement is lack of a valid system for valuing it across the many purposes and points of view that must be balanced. These problems can be explained fairly easily, but resolving them remains a central issue in our mixed economy and democratic political system.

Market and nonmarket value

One issue is market and nonmarket value. In the market, value means willingness to pay. However, what if you esteem something, such as clean water in a nearby stream, and consider it to have value even if you are not faced with a decision to pay for it or not? It might be that a lot of people esteem the same thing and want the political system to ensure that it remains, but do not want to pay for it. The nonmarket value controls and the political and legal systems must provide the mechanism to provide the clean water.

The *market value* of a good is what someone pays willingly for an item because in her mind the item's value exceeds its cost. Through application of such principles, the market provides an "invisible hand" that is said to meet all of our needs.[10]

Public goods, such as the clean stream water, do not obey the market economy, however. Water is one of those public goods which, as a commodity, do not obey market rules. In a public-good setting, common goods such as instream water are valued by individuals, but there is no direct mechanism for individuals to pay for them. For the most part, you do not choose between water and some other consumer good; you have to have water. Now, however, we are seeing some *unbundling* of water services. For example, people pay high prices for bottled water to drink, but assign lower values to larger quantities of water such as is needed to irrigate lawns.

Valuing water is hard because we lack a valid system to do it

Utility monopolies

A final issue in water economics is the incentives of the water providers. Water is a utility good that is usually provided by a monopoly provider, such as a local water department, and there is no competition among providers. When you do not choose between alternative providers, how should the monopoly provider decide what to charge? To bring safe water to consumers, the cost to divert, treat, and distribute water does not always represent its full cost because the *opportunity cost* of using water in alternative uses (such as environmental) may not be recognized. In other words, the market might allocate water for economic uses but ignore noneconomic ones.

Some water uses can be charged for because they are utility services. A public utility provides a necessary service or commodity, such as electric

---

10 The invisible hand is a widely quoted metaphor from eighteenth-century economist Adam Smith to explain how the free market works.

power, drinking water, or public transport. The market does not set the prices for these because of lack of competition, therefore some nonmarket mechanism such as the political process or a regulatory agency must be used to control prices.

Other water services involve public goods that differ from utility services. The criteria to identify these include the extent of filling public purposes, commoditization (can it be measured, unbundled, should it be rationed), and whether it is a natural monopoly or not.

When utilities are monopolies, the prevailing way to set their prices is through rate regulation, mainly through public utility commissions. However, most water utilities in the United States are publicly owned and not subject to such regulation. In effect, regulation of their rates is by local politics.

Summary points

- Getting people to value water fully is a key challenge for TWM, both to determine who will pay and to prevent waste. AWWA recognizes the importance of valuing water. Its intent to engage society in a discussion of the value of water is consistent with its focus on TWM.
- Replacing *good* with *value* in the opening statement of the TWM definition about the "greatest good" would help to bring economics into TWM more directly. The often vague and abstract language of water economics is a barrier to using it in more than just a theoretical way. The concept of the value of water is difficult to use in decision-making because of a lack of a valid system for valuing it across the many purposes and points of view.
- The best allocation of water returns the highest value to society, and this requires that values be assigned to the uses of water for drinking, irrigation, recreation, industrial production, and other purposes. However, people weigh these values differently, and that is why the political process is used in water planning.
- In a perfect world, society would balance the allocation of water by applying economic theory and a pricing system, but in practice things are worked out with money, politics, and the legal process.
- Our regulated and political system is one in which utilities select least-cost options and set rates at cost of service, with regulation of monopoly providers being provided by elected officials.
- Each decision-making body pursues its own self-interest, and the "it's not my problem" syndrome results from the gaps between the interests. Hopefully, decisions consider the Triple Bottom Line of economic, environmental, and social values.

- Economics is used for analysis and decision-making in water planning, whereas finance is a focused management tool that deals with how to pay for things. Economics uses tools such as benefit–cost analysis and environmental impact statements, whereas finance uses budgets, revenue and cost analysis, and financial statements.
- The theory of benefit–cost accounting is that for any proposed action the positive impacts (benefits) and negative impacts (costs) are estimated and compared. Guidelines for how these impacts are estimated must be based on standard methods, such as those developed by the US Water Resources Council.
- Benefits of water uses are computed differently for different uses. Examples are municipal and industrial water supply, irrigation, hydropower, navigation, and recreation. Flood damage reduction from a water project is also a benefit of water management, but it focuses on prevention of damage rather than on supplying a commodity.
- The balancing feature of TWM is that it allocates water effectively so that costs are commensurate with all benefits and values. The value of water addresses the benefit side of the equation, and the other side is how the costs are managed.

Review questions

1. Among the water uses, which are people most willing to pay for? Least willing to pay for? Explain these public choices.

2. Explain any differences in the terms "greatest good to society" and "value of water to society."

3. How does society go about determining the best allocation of water that yields the highest value to society? How does it take into account the fact that people weigh values differently?

4. What are the mechanisms by which the regulatory process is carried out for water allocation and uses?

5. Name economic tools used in water planning and tools used for financial planning.

6. Explain how the benefits for water used as municipal supply would be computed.

References

American Water Works Association. 2005. *Strategic Plan.* Adopted by the board on January 16, 2005.

Howe, Charles W. 1971. Benefit–Cost Analysis for Water System Planning. *Water Resources Monograph* 2. Washington, D.C.: American Geophysical Union.

James, L.D., and R.R. Lee. 1971. *Economics of Water Resources Planning.* New York: McGraw-Hill.

Mehan, G. Tracy III. 2007. God Gave Us the Water, but Who Pays for the Pipes? *Water & Wastes Digest* 47 no. 5 (May): 1.

Raucher, R., D. Chapman, J. Henderson, M. Hagenstad, J. Rice, J. Goldstein, A. Huber-Lee, W. DeOreo, P. Mayer, B. Hurd, R. Linsky, E. Means, and M. Renwick. 2005. *The Value of Water: Concepts, Estimates, and Applications for Water Managers.* Denver, Colo.: AwwaRF.

Water Resources Council. 1983. *Economic and Environmental Principles and Guidelines for Water and Related Land Resources Implementation Studies.* Washington, D.C.: Water Resources Council.

Young, R.A. 1996. *Measuring Economic Benefits for Water Investments and Policies.* World Bank Technical Paper No. 338. Washington, D.C.: World Bank.

CHAPTER 7

ENVIRONMENTAL WATER: ASSESSMENT, VALUE, AND SUSTAINABILITY

The fact that the market does not give a full picture of the environmental value of water is a big obstacle to TWM and has led to a lot of environmental regulations. In reality, this means that trade-offs are not occurring so much as regulations are being met. However, people feel sometimes that more balance is needed. The concept of value would take care of the problem if environmental resources could be fully valued, but they cannot. This chapter explains how we can account for environmental benefits and costs so that the Triple Bottom Line can indicate our commitment to the sustainable use of water.

Sustainability is a core element of TWM: "A basic principle of Total Water Management is that the supply . . . should be managed on a sustainable use basis." TWM also calls for stewardship, the greatest good, and a balance between society and the environment. It also "encourages planning and management on a natural water systems basis," and it "balances competing uses of water through efficient allocation that addresses . . . environmental benefits and costs." In short, sustainability and environmental use of water permeate the TWM philosophy.

This chapter begins with explanations of sustainable development and natural systems and then looks at the state of the environment today. Then it outlines environmental requirements for water. It cites

At the end of the day, TWM is about sustainability

problems and actions that affect the environment, using the four categories of chapter 2 (diversions, discharges, hydrologic modifications, and nonpoint sources). Then the impacts of these on natural systems (land, water, and species) are explained, as well as how these effects are monitored and assessed with reports according to a system of environmental indicators.

Sustainable development and natural systems

Chapter 1 outlined the general nature of sustainable development and sustainability, which are the notions of using environmental resources only to their carrying capacity so that resources are left for tomorrow. These notions have received wide acceptance around the world as policy goals, but there are often wide gaps between the rhetoric of acceptance and placing them into action. At any rate, the challenges are clear. World population and economic production are rising dramatically, and environmental resources are under more pressure all the time (Figure 7-1, based on data from Population Reference Bureau, 2007).

Managing water on a natural-systems basis is a companion notion to sustainability. Ideally, it requires that systems of land, plants, and living things be sustained as they were in their natural states before they had to compete with humans for water resources. Naturally, this ideal is difficult to achieve in altered ecosystems, but it gives us a target to shoot for. Managing water within watersheds offers us a way to balance uses of water and ecological systems within a natural accounting unit for water production and use.

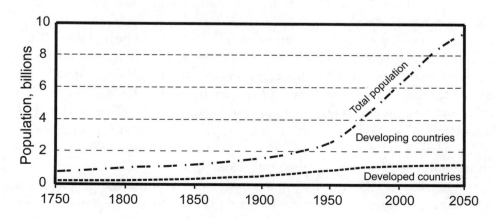

Figure 7-1.   World population with projection to 2050

In explaining watershed management, I have searched for the best illustrations and, although many are available, the one shown in Figure 7-2 seems to me to be the clearest and most complete. Amazingly, the drawing is around 60 years old and appeared in the frontispiece of the report of the President's Water Resources Policy Commission (1950).

The illustration shows the watershed as a source of water and its features for water utilization. The watershed is a prominent feature, and you can see two catchments, the larger one on the left with snowmelt feeding the reservoir, and the smaller one on the right with a small tributary stream. Several dams and reservoirs are shown. Most obvious is the large multiple-purpose reservoir. Just below is a diversion dam that enables the high line canal to take irrigation water from the stream. At the upper right is a beaver dam, and lower down is a regulating basin. On the main stem of the river is a re-regulating reservoir with a lockage system for navigation. A number of infrastructure components are also shown, including levees, outlet works, pipelines, irrigation systems, treatment plants, and pumping stations.

Sustainability is based on the concept of a healthy environment requiring a holistic approach to the needs of land, water, and living things. Land, water, and air have natural functions that enable living things to survive, prosper, and fill their biological roles in the food chain. If these natural functions are to work, the biosphere must sustain them at the same time that society's economic and social needs are met. In a perfect world, we would impose only light burdens such that natural renewal and replenishment of environmental resources occurred. However, this is not a perfect world, and our need to balance human and environmental needs requires compromises. Finding ways to make these compromises is the goal of environmental policy.

A balance is required because environmentalists worry that the compromises will degrade the environment, and economic developers worry that they will cost too much and lower our standard of living.

State of the environment

If there ever were a watershed year in environmental protection, it was 1970, when the first Earth Day was observed. This followed a series of 1960s events that focused attention on the environment, such as when we were able for the first time to view photos of the Earth from outer space. Although some good news has emerged since then, there has been mostly a rising tide of gloomy news about the state of the environment. At least the situation is getting a lot of attention.

TWM can be a powerful tool to affect important environmental problems. Its focus on sustainable water management is a visionary

164 TOTAL WATER MANAGEMENT

Source: President's Water Resources Policy Commission, 1950

Figure 7-2. Watershed showing natural and human systems

concept, but the real issues are in the details of water management, where situations are complex and people disagree on measures to take. To even discuss the issue, we need a clear picture. We begin with a picture shown in Figure 7-3. Here you see how actions under TWM have impacts on land, water, and living things.

The figure shows how TWM involves taking or regulating actions, detecting impacts by monitoring, assessing the impacts, and then making adaptive management decisions.

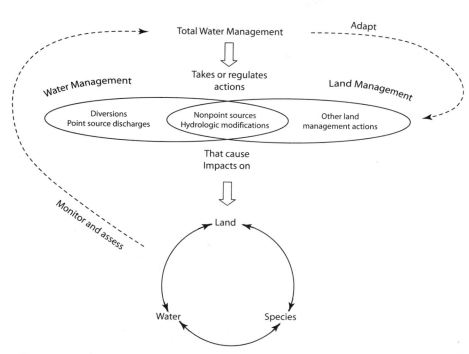

Figure 7-3. TWM as a comprehensive approach to water management

Which problems are most serious?

Given the rhetoric and noise level about the environment, it is hard to know which problems are most serious. Some 15 years after Earth Day, I reviewed the "water crisis" as it was being reported in the media and by government policy studies (Grigg, 1985). Based on these reports, serious problems were dam safety, flooding and stormwater, groundwater, polluted water, safe drinking water, and water scarcity. Other problems in the headlines seemed political, like the effectiveness of federal water planning, equity of cost allocation, financing problems, high water bills, interbasin transfer, and slow project completion. Of these problems, the one that has retained the highest profile is water scarcity.

The other issue with a high profile is water for the environment. Attention to this problem built up during the 1970s and 1980s and, just after the 1992 election, the Longs Peak Working Group (1992) issued a policy report on it.[1] They emphasized environmental sustainability gaps,

---

[1] The Longs Peak Working Group was an ad hoc group of experts who met in 1992 under the auspices of the Natural Resources Law Center at the University of Colorado to draft a set of policy proposals.

including endangered salmon, an overtaxed San Francisco Bay Delta, poison in the Kesterson National Wildlife Refuge in California, salt levels of the Colorado River, a vanishing Ogalalla Aquifer in the central plains region, Louisiana's eroding Delta, and the dying Florida Everglades. The working group used adjectives such as "endangered," "overtaxed," "poisoned," "salt-choked," "vanishing," "eroding," and "dying" to make their point that unless something was done, the outlook was bleak. These problems are still with us, but their long durations and the difficulty in their solution have dulled our sensitivity to them. War and global issues have also drawn the public's attention away.

So is the environment threatened or not? To get more insight I looked for credible assessments based on actual data. As of this writing, the issue that trumps the others is global climate change.

Global climate change is now at the top of our list of environmental concerns

If global climate change resulting from global warming is real, and most scientists seem to think it is, then the consequences to water managers will be dramatic. These might include more droughts in some cases and more flooding in others. They might include sea-level rise and inundation of coastal aquifers. They might include changes in the timing of runoff, the melting of glaciers, loss of natural water storage, and many other effects.

Regardless of energy use and global warming, climate varies over time naturally. Just because average conditions in a place have been about the same for many years does not mean they will stay at that level. Concern about greenhouse gases and the rise in CO^2 emissions comes at a time when natural climate changes may be occurring as well.

TWM offers us a way to respond to water changes caused by global warming by balancing and rebalancing water uses. The water-related activities of most concern are energy use and the release of greenhouse gases, which have been implicated as the probable causes of global warming. These have to do with how we live and use energy, and these root behavioral issues also affect the sustainability of water. To see the general patterns of water use and its effects on the environment, we turn to environmental assessment reports.

USEPA's report on the environment

The US Environmental Protection Agency (2003a) published its first national Draft Report on the Environment in 2003, using available data to answer questions pertaining to national environmental and human health conditions. The report uses environmental indicators to give scores to current conditions, trends, and data gaps. This is a summary of its key findings, with a focus on our subject, TWM.

To create this report, USEPA had to organize categories of environmental impacts and indicators for how each category was faring. In its view, some problems seem to have stabilized while others are difficult to assess. Stabilized problems are the condition of streams (where we lack a comprehensive picture), estuary conditions (poor to fair), the rate of annual wetlands loss, and chemical impacts on water. Problems that seem to be worsening are beach closings, fish advisories, and land development impacts. Ecological conditions seemed unclear because of difficulty in assessment.

Table 7-1 gives a summary of the report's conclusions in key areas relating to TWM.

Fish and Wildlife Service information

Another source of environmental information is from the US Fish and Wildlife Service (2006), which publishes information about species and ecology. Their focus is on species at risk, and they report that nearly 50 percent of the species at risk (20 percent endangered, 30 percent threatened) are water dependent because they either live in water, live in water during at least one life stage, or depend on aquatic plants and animals for food.

International reports

International institutions such as the United Nations (UN) and the World Bank also publish reports on the state of the environment. UNESCO's (2007) World Water Assessment Program's (WWAP) reports are aimed at developing the tools and skills to understand basic processes, practices, and policies to improve supply and quality of freshwater. As part of that process, the WWAP develops indicators to measure progress toward sustainable use of water. They have identified a number of challenges that focus on water use and increasing demands, and they report stress across sectors that include health, ecosystems, cities, food, industry, and energy. Their report focuses more on population and meeting basic needs and does not dwell on environmental issues.

The World Bank (2007) issues annual development reports that include indicators in six categories: worldview, people, environment, economy, states and markets, and global links. It includes more than 800 indicators but does not report an overall assessment of the global environment. In the case of water, a typical indicator is "improved water source," meaning the percent of population with access to water.

Table 7-1. Conclusions of the 2003 USEPA report

Issue	Conclusions
Waters and watersheds	We lack a comprehensive picture of the condition of waters at the national level. (We do have a compiled summary of water quality trends in the USEPA 303[d] report.) The nation's estuaries are in poor to fair condition, and the annual rates of wetlands loss have decreased.
Drinking water	An increasing number of people are served by Community Water Systems that meet health standards.
Recreation	There has been an increasing number of beach closings (this may be because of better monitoring).
Fish and shellfish	The percentage of freshwater under fish advisories has increased.
Land use	The amount and rate of land use have increased.
Chemicals	Industrial toxic releases have decreased and the nation is making progress in dealing with hazardous wastes.
Health	Human health is improving.
Ecological conditions	This is a difficult category to assess and USEPA does not have a good overall national picture.

Source: USEPA 2003a.

Summary of environmental issues

From the many available reports on the state of the environment, it appears that some environmental problems are getting worse and some are stabilizing. In other words, there is no clear picture. Global climate change, deforestation, and loss of habitat seem like inevitable consequences of world growth. Within this large-scale and difficult policy area, many environmental water problems call for our attention. The most visible ones are summarized in the next section.

Urban demands and land development

Around the globe, including in the United States, land development and urbanization continue relentlessly. In fast-developing nations like China and India, the pace is quicker as they develop their economies. Even in poor countries, growing populations cause indigenous people to clear and develop land, thus removing many beneficial environmental elements. The conflicts between development and environment will continue to be relentless and ratcheting upward. Much of the impact will be felt in the vulnerable headwaters regions.

River and stream impacts and pollution

Rivers and steams are the receiving waters for environmental impacts of land development and use. Some problems, such as pollution of rivers, seem stable, at least in the United States, but problems of biodiversity, fish disease, and loss of habitat seem to be taking us toward an unsustainable future. While large and visible pollution is less in the United States than before the Clean Water Act, we do not know how stream biology is changing as a result of other effects such as chemicals, pharmaceuticals, and nonpoint sources, which may be ratcheting stream integrity downward.

Chapter 10 explains how a program called Water Quality 2000 (1992) assessed the real causes of stream pollution. The impacts on water quality identified are outlined in Table 7-2. The top sources of environmental water impairment listed in USEPA's (2003b) 303(d) list included familiar pollutants such as sediments, pathogens, metals, nutrients, pesticides, and other common pollutants. The list differs somewhat for rivers and streams, lakes and reservoirs, and estuaries. Each state reports to USEPA under Section 305(b) of the Clean Water Act, and USEPA publishes the national summary.

Some environmental conditions have stabilized, but there is no clear picture on others

Coastal waters

Our vulnerable coastal waters remain of great concern. Whereas in the United States a good bit of attention has been given to estuary and coastal water programs, many problems remain, such as dead spots in ocean waters, as in the Gulf of Mexico off of Louisiana.

Fish, wildlife, and living things

As USEPA learned, it is hard to get a good picture of the viability of fish, wildlife, and living things. Some species seem to be thriving and adapting well. Urban wildlife, such as squirrels and some birds, seem to fall into that category. In parts of the country, with modern wildlife management, species such as turkeys and deer have increased. Fish will recover when conditions are improved through fishery management. However, disturbing indications are on the horizon that some species are under too much stress and that the diversity of species is threatened.

Lakes

One of the most obvious environmental issues related to water is threats to lakes and inland seas. We already explained how threatened estuaries and inland lakes of all sizes are being threatened by eutrophication and choking by algae, which is difficult to reverse. Inland salt lakes, like the Salton Sea and even the Dead Sea, have dropped in water level, thus threatening the regional ecologies that depend on them. The Aral Sea in Central Asia is an environmental disaster caused by diverting too much

Table 7-2. Impacts on water quality as defined in Water Quality 2000

Source	Impacts
Agriculture	Discharges sediment and nutrients along with smaller quantities of toxic chemicals. Accounts for wetland losses and damage to riparian and floodplain environments. Runoff from animal production is a source of phosphorus and pathogens in lakes, and agricultural chemicals threaten groundwater.
Atmospheric sources	Acidic or toxic substances may be deposited in lakes or estuaries. This may impair aquatic ecosystems, cause algal blooms, and even be lethal to aquatic organisms.
Community wastewater systems	Treatment plants remove much contamination, but they may not work well or remove toxics. They miss nonpoint sources and may be bypassed by combined sewer overflows.
Industrial dischargers	While industries are generally in compliance with permits, they discharge a massive quantity of conventional and toxic substances and thermal pollution.
Land alteration	Logging, mining, grazing, and land development change runoff and add sediment and chemicals to the water. They may also destroy wetlands and habitat.
Fish culture	Fish stocking and harvesting of aquatic species may impact aquatic ecosystems.
Transportation systems	Ships, roads, rail, and pipelines impact the waters. Oil spills are a major source of contamination. Transportation may destroy habitat as, for example, through dredging.
Urban runoff	As in land development, urban runoff causes contamination through release of sediment, organics, oil, and toxic chemicals.
Water projects	Water projects may reduce habitat through channelization, dams, and consumptive use of water, impacting anadromous and riverine fishes.

Source: Water Quality 2000, 1992.

water for upstream use, with the result being a drastic lowering of lake levels and shrinking shorelines.

Health risks

In addition to problems that are caused by overstressing environmental systems, including depriving these systems of water, humans are adding compounds of emerging concern, including endocrine disruptors and chemicals. We are still learning about the effects of prescription drugs, but the rise in their use alerts us to the fact that many used and expired drugs are being dumped into receiving waters, causing trace effects.

Conclusions about the environmental crisis

The environmental crisis appears to be real, but it will look different to the "environmentalists," the "water managers," and the "citizens." Measuring the impacts of water management actions on the environment requires better indicators than we have now. Clearly, pollution and pollutants of emerging concern are big threats. Species diversity appears to be decreasing. While a big focus is on climate change, environmental change is really like a problem out of Swift's *Gulliver's Travels*. Like Gulliver with the Lilliputians, myriad small impacts accumulate to create large impacts locally and globally.

> The environmental crisis is real, but it does not look the same to everyone

What are the water needs of natural systems?

Regardless of the state of the environment, it needs water to nourish it toward its natural state, even if its aquatic and terrestrial communities have been altered. People do not always agree whether the current state of the environment is satisfactory or not. We cannot answer that question here, but we can advocate that if the current environment condition is satisfactory or nearly so, then the water it took to establish it is a fairly good measure of ongoing water needs.

Definition of natural water systems

Interpreting the main goal of sustainable development—to preserve today's resources for tomorrow's generations—means that we do not want the environment to be degraded. We want to maintain it in a condition that is at least as good as it is in now. We are fortunate if it is close to natural conditions, but if it is not, we try to prevent deterioration or, in many cases, to try to improve the environment. Many battles are fought over this issue, such as the "zero net loss of wetlands" battle.

Having said those things, we can envision a range of environmental conditions from natural to degraded, with various intermediate states with altered but relatively good environmental conditions. Within these intermediate zones there would be environmental assets—such as clean water, wetlands, and habitat—and there would be liabilities in the form of too much development, pollution, and the like. To sum up, when managing the environment, we try to sustain or improve things from wherever we start. This requires water for terrestrial ecosystems and for aquatic ecosystems.

A starting point is to define what is meant by *natural systems*. A natural water system would be in pre-development or "virgin" condition.[2] It would not have any constructed works, diversions, or sources of contamination. Natural systems can, of course, be swamps and wetlands, high plains, desert, or any other ecological type.

> The natural system is a good benchmark, but most watersheds have some development in them

Water needs of natural system elements

It is convenient to consider the natural stage as a benchmark of what a water system can be and how it functions among other natural systems. Functions of natural water systems can be studied by viewing the hydrologic cycle from top to bottom. The hydrologic cycle is dynamic, with water quantity and quality changing constantly in its atmospheric water, surface water, and groundwater parts. As an environmental system, it includes water flows from atmospheric water all the way to flow into the oceans. In between, you have watershed runoff from land, river networks, riparian areas, lakes and reservoirs, wetlands, groundwater systems, and estuaries. One way to discuss the water needs of *natural systems* is in terms of these elements.

Watersheds

The watershed (drainage basin, catchment, or river basin) is a key component in TWM, which encourages planning and management on the basis of natural systems. A watershed is the land area draining to a point on a stream and is nature's production unit for water supplies. Watersheds are important accounting units for water resources management, and water is stored, filtered, and transported within them.

The watershed is an important source of drinking water. The TWM definition states that it "promotes . . . source protection . . . to enhance water quality and quantity." This recognizes the value of natural watersheds as sources of pure water, a principle that has remained a cornerstone of drinking water policy.

Unfortunately, few watersheds are in pristine condition. When they are, water quality and quantity are protected from human-caused threats

---

[2] The terms *virgin* water system or *natural flows* are used by hydrologists to describe what a system was like before water development occurred.

from land use activities, but they are not protected from natural disasters such as mudslides, fire, avalanche, volcano eruption, or drought. They are also vulnerable to the transport of pollutants by air. That problem occurs now in Rocky Mountain National Park, which is mostly natural but suffers from airborne nitrogen pollution.

Land use threats in watersheds are nonpoint sources from urbanization, transportation, industrialization, waste disposal, farming, ranching, logging, construction, and mining. Agricultural sources are the largest category of pollutant by volume, including sediment, fertilizers, pesticides, and herbicides. Hydrologic modifications to streams also change natural systems. Given the large number of headwater streams, many of these modifications occur in them.

Poor watershed management is a major cause of land and water degradation and rural poverty in the world. Watershed management measures include regulatory instruments (zoning, regulations, land and water rights, controls, permits, prohibitions, and licenses); fiscal controls (prices, taxes, subsidies, fines, and grants); and direct public management measures (technical assistance, research, education, land management, installation of structures, and infrastructure). All of these measures can be employed in TWM.

Rivers and streams

The main channels of rivers and streams are bordered by floodplains created by geologic forces. The riparian corridor is the strip of channel and floodplain wetlands that sustains the aquatic ecosystem. Maintaining it in healthy condition is critical to the functioning of natural systems. So the water needs of rivers and streams comprise the needs of the channel itself and the riparian corridor. Preserving healthy conditions in the riparian corridor is a land use issue as well as a water issue. The quantity and quality of water in the streams themselves depends on success in management of instream flows and of the adjacent aquifers.

Because stream dynamics are complex, it is hard to describe their water needs. If the hydrologic regime is unaltered and the water clean, availability of habitat is ensured in the streams and adjacent wetlands and floodplains, but this is a difficult goal to attain.

Hydrologic alteration can occur with changed flow rates, schedules, and volumes. It can cause changes in flow regime, low flows, unwanted flooding, sedimentation and erosion, and habitat destruction. Water quality changes can include alterations in any of the natural substances that are historically found in a stream, and they can lead to toxic substances in the water.

The issue with instream flows is to maintain adequate flow quantity and quality in the stream at all times both for all intended uses and for the natural systems. TWM should also address water quality by promoting source protection and supply development for both water quality and public health. Water quality is like art: you may recognize good and bad water quality, but it is difficult to measure and reach agreement about it. The Clean Water Act regulates stream water quality and sets standards for it.

A minimum instream flow quantity is necessary to maintain stream water quality, but even more water may be required for fish and wildlife. Determining the needs for water quantity in the stream is an important and contentious issue. Whereas the Clean Water Act establishes rules for water quality, no one really establishes rules for instream flow quantities.

The instream needs include water carriage needs. Instream flows for withdrawal uses are straightforward to analyze because they are a matter of quantity and timing of water flow. Environmental uses are more difficult to quantify, however, because of biological systems, species life cycles, and other complexities (Waddle, 1992).

Instream flow decisions consider physical, legal, hydrological, and ecological conditions. They should use fish life cycle requirements to set operational requirements and aquatic system needs. Viewing instream flow requirements as a static minimum flow may be a mistake as variation of flows is also required.

Many complexities enter the picture, including species and life stages of fish; floods; water quality; and fishing pressure. Hydrologic-biological methods to determine flow requirements are available. The Fish and Wildlife Service sponsors a method called the Instream Flow Incremental Methodology (IFIM), which includes hydraulics, channel structure, hydrology, water quality, and micro- and macro-habitat elements. The methodology has a program within it called the Physical Habitat Simulation System (PHABSIM).

Other methods including commercial software are also available. The Nature Conservancy sponsors a computer program to estimate hydrologic alteration and compute how much water a stream needs (Richter et al., 1997).

Lakes and reservoirs

Reservoirs and lakes are also important parts of the natural flow regime, even if they are man-made. An unregulated lake can simply be part of the stream system, and if it is small enough, it can be practically part of the stream. In that sense, its water needs depend on its size, location, and position relative to the stream network.

A regulated reservoir can smooth flood flow or provide stored water for downstream demands for water supply, irrigation, fish and wildlife, recreation, flood control, navigation, hydroelectric power, and water quality improvement. Releases from reservoirs determine instream flows in regulated streams according to guidelines and decisions based on rule curves or control center forecasts. Once such a reservoir is in place, its water needs become somewhat artificial, in the sense that without an adequate supply of water its ability to provide releases is limited.

> **Water storage is essential for its management, but artificial lakes alter natural flow regimes**

Lakes and reservoirs have different flow regimes than do streams. In the lakes, water flow is slowed down and its velocity is less. The water surface is subjected to a different heat budget, so its temperature will change, and at lower depths, the water temperature may remain colder because it is not exposed to the sun. Also, light does not penetrate as well to lower depths, and photosynthesis is altered. Lakes differ according to seasons, winds, and other forces. Lake turnover is an important aspect of annual cycles. When the heaviest water is on the lake bottom, it will remain there, but when it is on the surface, it will sink, causing a turnover.

Water quality changes from storage may affect low flows, oxygen, or movement of pollutants in a river reach. A reservoir's water environment is affected by currents, temperature, light, wind action, and other climatic conditions. Algae blooms in lakes are an important water quality problem. Lake sediments play important parts in water quality. Chemical and biological contaminants can become trapped in sediments and remain for many years.

The environmental cost of water storage is that it changes natural flow and alters stream corridors, with effects on water flows and water quality and altered conditions for habitat. Reservoir pool management is important for lake fishes and waterfowl, as well as for the aquatic species affected by releases. The ecology of a reservoir will be different from the streams that supply and drain it, and an aquatic and terrestrial ecology will develop around it. Even small reservoirs have important cumulative effects in watersheds.

Wetlands

Wetlands have many valuable functions. The definition of a wetland tends to be technical, but they include swamps, marshes, bogs, sloughs, potholes, wet meadows, river overflows, mudflats, and natural ponds. Freshwater marshes have diverse kinds of grasses, whereas swamps are often dry in summer and may be characterized by woody plants including trees. Saltwater marshes and swamps serve as habitat areas for a wide variety of saltwater fish and coastal wildlife.

Wetlands are sustained by rainfall, springs, or floodplain flow. They are feeding, spawning, and nursery grounds for more than half the saltwater finfish and shellfish harvested in the United States annually, and most of the freshwater game fish. They constitute habitat for a third of the resident bird species, more than half the migratory birds, and for many endangered and threatened plants and animals. Wetlands also lock up peat and prevent it from being discharged into the atmosphere.

Wetlands are valuable components of water systems

Environmental functions of wetlands include providing habitat for fish, birds, and other wildlife; protecting groundwater supplies; purifying surface water by filtration and natural processes; controlling erosion; providing storage and buffering for flood control; and providing sites for recreation, education, scientific studies, and scenic viewing. Wetlands face natural catastrophes from flood, drought, ice damage, high winds, waves, and fire. They can buffer ecological systems from damage due to these catastrophes, although not without stress (Grigg, 1996).

Groundwater systems

Groundwater is a dynamic part of the hydrologic cycle, just like surface water. The main differences are that groundwater moves much slower and is exposed to different chemical and biological environments. Some groundwater, called fossil water, may have been in storage for thousands and even millions of years. Other groundwater, in tributary aquifers, may flow almost as quickly as surface waters.

Groundwater has been utilized from springs or tapped through wells that date to one or two millennia BCE. The only place you can see groundwater in its natural form is in limestone caverns or other large, subterranean openings.

Estuary functions

Estuaries, or water bodies formed by the confluence of a freshwater channel and the sea, have high biological productivity. Examples are found in coastal regions such as California's Bay-Delta system, Louisiana's Mississippi Delta, the Nile Delta, and the Chesapeake Bay. Ecological issues in estuaries include nutrient balances, freshwater inflows, grass and submerged aquatic vegetation, fisheries, benthos, and the food chain.

Estuaries˜ are threatened because many major cities lie next to them and many people depend on them for income and food. There are about 850 estuaries in the United States alone (National Academy of Science, 1983). Pressure on them is from population growth, agriculture, industrialization, fisheries, and disposal of dredge spoil. Spec-ific water problems in coastal areas include (Davies, 1985):

- Nutrient enrichment, eutrophication, and nuisance algae
- Threats to dissolved oxygen levels
- Shellfish bed closures
- Lost and altered wetlands
- Disappearance of submerged aquatic vegetation
- Threats to living resources from toxics
- Diseased fish and shifts of fish species
- Salinity intrusion
- Groundwater problems

Through the National Estuary Program, USEPA learned a great deal about collaborative, problem-solving approaches to balance conflicting uses while restoring or maintaining an estuary's environmental quality. USEPA created the concept of the Management Conference as an umbrella for action and cooperation to achieve consensus that leads to action. These are a good example of TWM tools.

Water management actions and impacts

The main water management actions that can be controlled by TWM were explained in chapter 2, including diversions, discharges, hydrologic modifications, and nonpoint source discharges. This list provides a few examples of these actions:

> **Estuaries provide valuable water and environmental functions, but they are vulnerable to development pressures**

- Initiate or increase water diversion
- Discharge wastewater or change its volume or quality
- Construct a dam and reservoir
- Operate or change releases from a reservoir
- Construct a road and culvert or bridge
- Change land cover and its rate and quality of runoff
- Pump groundwater or change the rate of pumping
- Modify natural stream channels
- Apply fertilizer and/or chemicals to farmland
- Drain wetlands or change their hydrologic character
- Do construction or maintenance that causes erosion and sedimentation
- Change irrigation and drainage patterns

- Conduct fish farming operations
- Conduct forestry operations that change watersheds

The players who take these water management actions range from large utilities to small landowners. A list of them would include: utilities; road departments; agencies with dams, canals, and water infrastructure; agencies that develop land; industries with self-supplied water services; farmers; anyone who dredges a stream; and the owners of ponds and reservoirs. In addition to these owners, regulators and the consultants and designers who specify developments can be said to take water management actions.

Impacts on natural systems

The impacts of these water management actions can be measured in terms of loss of water through diversions, altered quality and volumes through discharges and nonpoint source discharges, and changed water ecologies through hydrologic modifications. These effects are interrelated through their integrated effects on species.

For example, stream ecology and life are complex, and the river continuum concept is sometimes used to explain how rivers change through different ecologic zones (Cushing, 2002). As water is pulled downhill by gravity, streams form dynamic systems of water, earth, and biological communities. As nutrients are dissolved from rocks by flowing water, they nourish algae and add to the food web with more and greater varieties of insects. Steam biological energy is driven by photosynthesis and the production of algae, which also contribute to the food web. Light is a critical element here to stimulate the growth of algae, moss, and plant material in streams. Clear streams with light penetration are more likely to grow plant material than cloudy streams, if other necessary elements are present. If these processes are disrupted by diversions, discharges of pollutants, sedimentation, or hydrologic alteration, there will be changes in stream production, fish and wildlife, and water conditions. The changes may at first be gradual, but a ratchet effect may alter the aquatic ecosystem forever.

> **Small actions change water regimes through a ratchet effect**

Richter et al. (1996) examined the functions of aquatic, wetland, and riparian ecosystems that depend largely on hydrologic regimes, with a focus on the variation in conditions that helps sustain the viability of species. They argued that modifications of hydrologic regimes can indirectly alter aquatic ecosystems through effects on habitat and water quality and quantity. They proposed a group of some 32 hydrologic attributes to capture the relevant variables. These were grouped into five categories to capture the magnitude of monthly water conditions, magnitude and duration of annual extreme water conditions, timing of annual extreme water conditions, frequency and duration of high and low pulses, and rate and frequency of water condition changes.

Balancing environmental benefits and costs in TWM

If we are trying to improve the environment while balancing "competing uses of water through efficient allocation that addresses . . . environmental benefits and costs," we face a relative situation that requires the valuing of these competing uses.

The main reason that it is difficult to reach consensus about environmental trade-offs is that people value them differently. A second and also important reason is that often science does not tell us clearly what the costs and benefits will really be.

We discuss the valuing environmental benefits in a graduate water management class at Colorado State University and use an editorial from the *Wall Street Journal* (1991) to illustrate the issues involved. The writers tend to be conservative and favor business solutions to environmental problems, so I explain that to the students in an attempt to provide a balanced view.

In this case, the writers were complaining that species in the Northwest were given "sacrosanct status" to the detriment of practical economics. They reviewed the issue of restoring salmon runs in the rivers by taking water management actions that involve utilities and farmers. They identified as a problem that "Litigious environmentalists will swim upstream as far as their bankrolls carry them to get rulings based on the findings of some apocalyptic scientist who doesn't have to consider the jobs at a pulp mill or an aluminum plant or the pollution from a coal plant that might replace hydro."

They expressed pessimism that the market would be allowed to make the decisions: "These days, when resource trade-offs prove too tough for legislatures of federal agencies to make, U.S. judges step in." They favored the market as the preferred decision-making mechanism: "Approximating a free market in natural resources isn't going to be easy—especially when so many parties have careers and causes at stake. But it's hard to think of any other mechanism capable of arbitrating the myriad demands of millions of people in an economy." However, in the end they remained pessimistic: "We're

> **Water markets offer attractive possibilities, but they are not able to allocate water to all needs**

not much closer politically to making the proper trade-offs today than we were when the big dams went up early in this century."

It is easy to give the students the other side of the story because, as chapter 6 explained, the market's ability to allocate resources this way is very limited. TWM provides a coordination mechanism to enable us to go beyond the market and not necessarily with the intervention of legislatures, federal agencies, or judges. To provide the coordination, we must find out what are attractive environmental benefits, what the costs are, what the public wants and will pay for, and what the legal requirements are. Then the package is assembled and the environmental benefits and costs are weighed against each other.

Examples of weighing environmental benefits and costs

Tampa Bay's program

In chapter 6, Tampa Bay Water was cited as an example of how a decision model was used in a process of shared governance to make trade-offs in the selection of water sources (Tampa Bay Water and CH2M HILL, 2006). The multiattribute analysis tool, called the Source Management and Rotation Technology Tool, enabled the utility to work with stakeholders to find optimum solutions for the selection of water supply sources. The optimum is defined in terms of both environmental and economic factors.

In this case, Tampa Bay's environmental issues center on groundwater pumping, which can cause loss of habitat and even land subsidence. Economic issues center on the cost of pumping, reliability, and related water supply operating questions. Balancing environmental benefits and costs requires a valuing the of the relative merits of the different options for source rotation, and Tampa Bay's model places everything into a multiattribute model for display and decision-making by its board.

TMDL program

Another example is the Total Maximum Daily Load (TMDL) program, which was established by Section 303(d) of the Clean Water Act. The TMDLs specify how much of specific pollutants must be reduced to meet water quality standards in a watershed, and then pollution control responsibilities are allocated among the sources in the watershed. TMDLs are supposed to integrate solutions to water quality problems from point and nonpoint sources.

The trade-offs here arise when, after identifying how much a given pollutant must be reduced, an allocation of responsibilities is made. The environmental benefits are decided on a command-and-control basis; that is, the level of environmental benefit is mandated by the Clean Water Act and subsequent regulations. The trade-offs occur when deciding how to meet the target. However, the trade-offs do not in this case involve reducing environmental quality.

Environmental monitoring and assessment

Monitoring and *assessment* are accepted technical terms for the processes of collecting data that tell us the state of an environmental system and assessing what it means. Environmental assessment is the process of assessing positive and negative impacts of a proposed action on natural systems. To perform an environmental assessment, you identify the environmental resources that are affected by an action and assess how they will be impacted. Figure 7-4 illustrates how assessment and decision-making are connected. The information platform serves to house data management programs, which go with the assessment theory and decision processes to create informed and integrated decisions.

Since the National Environmental Policy Act (NEPA) was passed, the nation has accumulated a great deal of experience with environmental assessment, which has been driven by the Environmental Impact Statement (EIS) process (see chapter 9).

Under the Clean Water Act, USEPA (2006) awards Section 106 program management funds to states only if they have adequate monitoring and assessment programs. The results must be provided in the state's Section 305(b) report. The monitoring programs should have the following elements:

- A comprehensive monitoring program strategy that addresses all state waters, including streams, rivers, lakes, reservoirs, estuaries, coastal areas, wetlands, and groundwater.

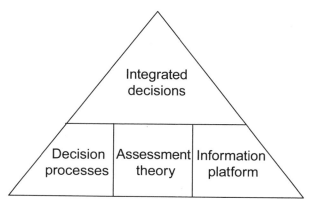

Figure 7-4.   How assessment and decision-making relate to each other

- Monitoring objectives that include but are not limited to Clean Water Act goals.
- Monitoring design to serve the state's monitoring objectives.
- Core and Supplemental Water Quality Indicators, such as indicators to represent designated uses. Core indicators include physical/habitat, chemical/toxicological, and biological/ecological data. Supplemental indicators are used for specific pollutants, when there is impairment, or to support special studies.
- Quality Assurance to ensure scientific validity of monitoring, laboratory, and reporting activities.
- Data Management for water quality, fish tissue, toxicity, sediment chemistry, habitat, and biological data.
- Data Analysis/Assessment based on various types of data (chemical, physical, biological, land use) from various sources, for all waterbody types and all state waters.
- Reporting under Sections 305(b), 303(d), 314, and 319 of the Clean Water Act and Section 406 of the Beaches Act.
- Programmatic Evaluation to determine how well the program serves its water quality decision needs.
- General Support and Infrastructure Planning to implement its monitoring program strategy, including funding, staff, training, laboratory resources, and improvements.

Assessment at the watershed level

These requirements provide an overall picture for the somewhat narrow purposes of managing the Clean Water Act programs at the river basin level, but many problems are at smaller scales and in headwaters regions. At the watershed level managers need better indicators to facilitate local decision-making. The problem is not a shortage of indicators. In fact, there may be a glut of indicators, with watershed planners and managers being data rich and knowledge poor. It is difficult to create indicators that serve the purpose, and many lists for multiple physical, social, and environmental purposes have been proposed. However, the lists are unfocused and not coordinated with the actual needs of watershed stakeholders and managers (UNESCO, 2003; United Nations, 1992).

Water resources assessment means a judgment of the state and condition of water resources. To assess means about the same thing as to appraise or to evaluate. It means to estimate the state, value, or condition of something.

To implement assessments, *indicators* are needed. They are signs or symptoms. Usually, the variables measure quantity, quality, availability, adequacy for the environment, and other parameters. Scale is one of the important issues to determine if the issue is at a watershed, river basin, state, or national level. It is difficult, however, to find standards against which to judge. What is "poor" water quality, for example? Unless some standard is set, the observations of the variables constitute "data" rather than "information." When you combine information with judgment that is based on a standard, then you are able to make an assessment of the state of the system compared to critical thresholds of impacts and consequences.

Valid decision information has value, because it can produce better decisions and higher-value returns. Conversely, too much information can confuse the picture and reduce its own value. So, both information and its careful organization add value to the decision process.

While assessment programs such as the UNESCO's (2003) World Water Assessment Program are impressive, tools for the large numbers of small watershed management groups remain inadequate, and in some cases there are no tools at all for localized cases. To illustrate, let's consider the actions of the some 3,000 watershed groups in the United States.

In Colorado, for example, the Colorado Watershed Assembly (2005) describes the groups this way: "In Colorado, there are more then forty local watershed initiatives involving citizen-based watershed groups. They vary in their missions and participants but all share a common focus: a strong commitment to protect the health of the aquatic systems and the life that depends on them. Often these groups are collections of existing groups that are interested in working on common problems through a consensus approach." Nationally, the groups meet at the annual meeting of the River Network (2004), a coordinating organization that focuses on local watershed solutions.

Valid water resources assessment is critical to management, but difficult to perform

The need for indicators was expressed at the 2001 National Watershed Forum (Meridian Institute, 2001). The nearly 500 participants representing diverse local watershed groups and other stakeholders confirmed that local citizens are forming partnerships to address problems affecting their water resources because, despite billions of dollars invested in reducing pollutants, many problems remain. They highlighted problems arising from top-down approaches, lack of information, lack of trust, lack of capacity, and other factors. Their recommendations focused on watershed planning and evaluation, data collection, monitoring, research needs, and information management. They called for improved access to information for stakeholders, a national campaign to highlight the importance and awareness of watershed issues, a clearinghouse for information about watershed protection, and a definition of a "healthy watershed" encompassing all elements and interrelationships.

More than 100 recommendations were made, many related to planning, assessment, and information management. A few of these were:

- Use common language to measure habitat and ecosystem functions
- Define the purposes of monitoring and correlate with decision-making
- Give unified messages to the public, watershed groups, and landowners
- Carry out broad-scale ecosystem assessments
- Clarify data to support growth decisions
- Use flow criteria for biological resources in water
- Incorporate sound science and source water assessment data
- Measure change in watersheds over time
- Recognize relationships between land, water, and ecosystems
- Use predictive models of consequences of development on land and water
- Ensure transparent and informed decision-making processes

The outcomes of the forum illustrate some of the institutional problems of watershed management. Watershed planning forums often founder from information overload, lack of conceptual clarity, and problems of data validity and reliability. The level of participant understanding of decision information varies widely. Unless language is readily understandable to all, time will be spent on explaining terms and concepts with continuing communication problems. Stakeholders encounter a wide variety of variables that require indicators, ranging across concepts such as water scarcity, water quality, and flood risk. In addition, different models must be used in combinations with critical variables connecting causes and effects. There is also a problem in translating general, policy-level indicators into the decision information required at local watersheds.

Much of the action in TWM occurs at the local watershed level, where valid information and capacity-building of participants are required

This latter problem is shown by Figure 7-5, which illustrates the macro and micro aspects of watersheds. At the macro level, the larger goals and policies are expressed at a high and aggregated level. At the micro level, smaller watersheds and local communities struggle with site-specific issues. A great deal of watershed planning and management

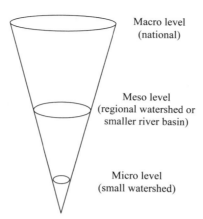

Figure 7-5. Micro and macro aspects of water planning

also occurs at the intermediate level, which might correspond to an intra-state river issue.

Theories about watershed planning tend to break down at local levels, preventing integrated decision-making. Lack of clear goals and information can confuse participants. Whether participants are paid workers or volunteers, they will participate only if clear issues, goals, and potential actions are presented. Without this clarity, expectations will be unrealistic and conflicts will be greater. Clarity depends on issue identification, goal setting, and assessment of the state of the watershed. These require a range of supporting indicators, such as water quantity, water quality, competing and conflicting uses of water, and agreements as to specific implementation steps.

As the focus shifts to higher levels, policy decisions depend on aggregations of the information from the watershed level. Unless information is aggregated by indicators that are based on common baselines and aggregated indices, the result can be a set of information that lacks both coherence and integrity, sending erroneous policy signals to decision makers.

The national watershed movement contains many examples of local experiences. Concerns are expressed, as for example in Southeastern states, about loss of biodiversity and how the losses accumulate into higher-order streams. Then the quality of water, sediment, and biota progressively deteriorate and result in macro-level effects like the dead zone at the mouth of the Mississippi River. Many of these effects go unnoticed because of limits in monitoring and assessment programs. In the western United States, the same trends are evident, but the water-based ecological conditions are different. This region has been characterized by rapid population growth, with impacts on watersheds often causing unforeseen

Table 7-3. Targeted watersheds and their features

Watershed	Size, sq mi	Features
Bear River, Utah, Idaho, and Wyoming	7,500	Integrated information system; water quality trading program; water quality modeling
Cape Fear River Basin, North Carolina	23% state area	TMDL program, water quality credit trading pilot, testing of a regulatory framework
Dungeness River, Washington	200	Microbial source tracking; BMPs for stormwater, septic maintenance, water treatment, and water conservation
Fourche Creek, Arkansas	170	Decrease hypoxia contaminants; improve wetlands and water quality; increase public involvement in urban watershed
Ipswich River, Massachusetts	155	Restore headwaters; quantify benefits of low-impact development (LID); implement and quantify water conservation
Kalamazoo River, Michigan	2,020	Model trading and framework for agricultural participation; address eutrophication issues through a phosphorus-based TMDL
Kenai River, Alaska	Unknown	Reduce hydrocarbon emissions; decrease use of 2-stroke boat motors; reduce effects of boat wakes on streambank erosion
Lake Tahoe, California and Nevada	Unknown	Water quality trading; removing fine sediments and nutrients in cold climates; numeric estimates; basin-wide nutrient reduction
Nashua River, Massachusetts and New Hampshire	132	Land protection; forestry; conservation; restoration projects; smart growth project; public survey; water quality sampling
Passaic River, New Jersey	669	Trading program focusing on point-to-point and point-to-nonpoint source trading; meet phosphorus-based TMDL for the river
Schuylkill River, Pennsylvania	130	Riparian buffer; process to remove phosphorus; reclaimed acid mine drainage discharge for power generation
Siuslaw River, Oregon	773	Restore landscape, incentives to reduce sediment to streams; restore habitat; protect estuary corridor; water quality monitoring
Upper Mississippi River, Iowa	Unknown	Reduce nitrate load to Gulf of Mexico by structural modifications to subsurface drainage systems; water quality monitoring
Upper Sangamon River, Illinois	Unknown	Reducing nutrient discharges; geographic information systems (GIS) and precision agriculture; drainage and subsurface bioreactors

Source: USEPA 2004.

and early warning signs, as for example in watersheds along Colorado's Front Range.

While watersheds receive financial support, much more is needed. USEPA's (2004) Targeted Watershed program offers grants as incentives to watershed organizations to implement clean water and watershed restoration efforts. Table 7-3 outlines the results of a review of watersheds identified for funding and demonstrates the range of scales, locations, and problems addressed by watershed programs. In these watersheds, impacts are felt from the local scale to the macro level of the entire Gulf of Mexico.

Summary points

- Sustainability and environmental uses of water are core issues in the practice of TWM.
- Sustainability is based on the concept of a healthy environment requiring a holistic approach to the needs of land, water, and living things.
- Managing by the watershed using TWM principles offers a way to balance uses of water and ecological systems within a natural accounting unit for water production and use.
- TWM's focus on sustainable water management is a visionary concept, but real situations are complex and people disagree on them. One issue that galvanizes the public's attention is water scarcity. Another issue with a high profile is water for the environment. Both of these illustrate why the political process is necessary in water resources planning and management.
- From the many reports available, there is no clear picture on the state of the environment. While some environmental problems are getting worse and some seem to be stabilizing, conflicts between development and environment will continue, and impacts will be heavy in vulnerable headwaters regions.
- Regardless of its state, the environment needs water to nourish it toward its natural state, even if its aquatic and terrestrial communities have been altered.
- The watershed is a key management unit in TWM, which encourages planning and management on the basis of natural systems. It is an important source of drinking water, and TWM promotes source protection to enhance water quality and quantity. This recognizes the value of natural watersheds as sources of pure water, a principle that has remained a cornerstone of drinking water policy.
- The main water management actions addressed by TWM are diversions, discharges, hydrologic modifications, and nonpoint source discharges.

These are controlled by players who range from large utilities to small landowners.
- Balancing competing uses of water through efficient allocation that addresses environmental benefits and costs requires the valuing of these competing uses. It is difficult to reach consensus about these values or to measure costs and benefits well.
- Monitoring and assessment require collecting data and assessing the state of an environmental system. To use TWM at the watershed level, managers need better indicators, which are signs or symptoms. Valid indicators have value, because they can produce better decisions and higher-value returns.
- In watershed planning, understandable language is required because communication problems among stakeholders makes consensus-building difficult and prevents integrated decision-making.

Review questions

1. Explain how managing by the watershed under TWM practices promotes a sustainable environment.

2. Using principles of TWM, explain how it responds to conditions of water scarcity and the needs of the environment for water supplies.

3. Explain how environmental water needs are determined.

4. How can the principle of protecting natural watersheds as sources of pure water be implemented?

5. How does TWM deal with the difficulty in reaching consensus about environmental benefits and costs of water actions?

6. Explain the differences in the terms *monitoring* and *assessment*. What are environmental indicators and how are they used in monitoring and assessment?

References

Colorado Watershed Assembly. 2005. http://www.coloradowater.org/. Accessed January 19, 2005.

Cushing, Cobert E., and J. David Allan. 2002. *Streams: Their Ecology and Life*. New York: Academic Press.

Davies, Tudor. 1985. Management Principles for Estuaries. US Environmental Protection Agency. Unpublished.

Grigg, Neil S. 1985. *Water Resources Planning*. New York: McGraw-Hill.

Grigg, Neil S. 1996. *Water Resources Management: Principles, Regulations, and Cases*. New York: McGraw-Hill.

Longs Peak Working Group. 1992. *America's Waters: A New Era of Sustainability.* Boulder, Colo.: Natural Resources Law Center, University of Colorado.

Meridian Institute. 2001. *Final Report of the National Watershed Forum.* June 27–July 1, Arlington, Virginia. Jericho, Vt.: Meridian Institute.

National Academy of Science. 1983. *Fundamental Research on Estuaries: The Importance of an Interdisciplinary Approach.* Washington, D.C.: National Academy Press.

Population Reference Bureau. 2007. Data Sheets. http://www.prb.org/. Accessed June 7, 2007.

President's Water Resources Policy Commission. 1950. *A Water Policy for the American People.* Washington, D.C.: Government Printing Office.

Richter, B.D., J.V. Baumgartner, J. Powell, and D.P. Braun. 1996. A Method for Assessing Hydrologic Alteration Within Ecosystems. *Conservation Biology* 10: 1163–1174.

Richter, B.D, J.V. Baumgartner, R. Wigington, and D.P. Braun. 1997. How Much Water Does a River Need? *Freshwater Biology* 37: 231–249.

River Network. 2004. http://www.rivernetwork.org/. Accessed January 19, 2005.

Tampa Bay Water and CH2M HILL. 2006. *Decision Process and Trade-Off Analysis Model for Supply Rotation and Planning.* AwwaRF Report. Denver, Colo.: AwwaRF.

United Nations. 1992. Protection of the Quality and Supply of Freshwater Resources: Application of Integrated Approaches to the Development, Management and Use of Water Resources. Chapter 18 of *Agenda 21.* New York: United Nations.

United Nations Educational, Scientific and Cultural Organization (UNESCO). 2003. *Water for People, Water for Life.* UN World Water Development Report. UN Economic Commission for Europe. http://www.unesco.org/water/wwap/wwdr/index.shtml. Accessed December 15, 2004.

UNESCO. 2007. *World Water Assessment Programme.* http://www.unesco.org/water/wwap/targets/index.shtml. Accessed May 21, 2007.

US Environmental Protection Agency. 2003a. *Draft Report on the Environment.* Washington, D.C.: USEPA.

USEPA. 2003b. *National 303(d) List Fact Sheet.* http://oaspub.epa.gov/waters/national_rept.control. Accessed December 10, 2003.

USEPA. 2004. *Targeted Watershed Grant Program.* http://www.epa.gov/owow/watershed/initiative/index.html. Accessed December 15, 2004.

USEPA. 2006. *Monitoring and Assessing Water Quality.* http://www.epa.gov/owow/monitoring/elements/execsum.html. Accessed October 26, 2006.

US Fish and Wildlife Service. Division of Environmental Quality. 2006. *Water Quality.* http://www.fws.gov/contaminants/Issues/WaterQuality.cfm. Accessed November 15, 2006.

Waddle, T.J. 1992. A Method for Instream Flow Water Management. Ph.D. Diss., Colorado State University, Fort Collins.

Wall Street Journal. 1991. Lox Horizons. May 20.

Water Quality 2000. 1992. *A National Water Agenda for the 21st Century.* Alexandria, Va.: Water Environment Federation.

World Bank. 2007. *World Development Indicators* 2005. http://web.worldbank.org/. June 2, 2007.

CHAPTER 8

SOCIAL IMPACTS OF WATER MANAGEMENT

The fact that environmental water needs are not fully valued by the market has led to our system of regulatory control and created glitches in the practice of TWM. It is also true that social needs are not fully considered by the market. On the social side, people need water services to survive and prosper, and these are not discretionary services to be bought only if you can afford them. Water and its uses are woven into the social fabrics of all societies.

The environmental side was discussed in the last chapter. TWM addresses social requirements for water in the general statements that it "addresses social values" and "fosters public health, safety, and community goodwill." This chapter expands on these concepts and explains how TWM works to take care of these social requirements of water.

The relationship of TWM to society goes both ways. Water actions have impacts on society and society has a responsibility to be stewards of water. Figure 8-1 illustrates the responsibility of the water industry to reach out to society to engage them in its actions.

These general requirements go along with other rights that we in North America expect, such as freedom, lack of discrimination, equal opportunity, justice, basic health care, access to basic services, and a chance to participate in the good life. While these are simple-sounding concepts, a lot of thinking has gone into them. For example, Derek Bok (1996), former President of Harvard, wrote *The State of the Nation: Government and the Quest for a Better Society* about what Americans want from life

192   TOTAL WATER MANAGEMENT

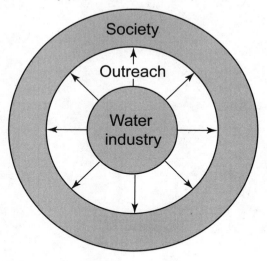

Figure 8-1.   Water industry and outreach to society

in their quest for a better society. Although people disagree about some social issues, most people still consider access to water services to be a basic human right.

So, if society has a water-related value, TWM addresses it by considering the social impacts of water-management actions. Although access to water is a basic human right, social aspects of water management can be controversial. Say, for a example, a water utility decides to add amenities to a water supply project, and includes fishing, boating, and daytime recreation on a reservoir. These additions to the project create an extra cost to the utility, which must be absorbed in its rate base. A group of citizens decides to sue because they claim this is taxation without representation, and besides, they do not care to use these facilities. Their claim is that recreational users must bear the full cost of the added amenities. While on the basis of pure rate setting this claim makes sense, the practical situation is that no recreation authority exists to provide the amenities, and even if it did, it would greatly increase the cost for the recreation, and the citizens would not be able to afford them. The other side of the argument is that water management inherently is a multiobjective activity, and it makes perfect sense that the recreation would be provided for and included in the rate base. Of course, there is no perfect answer and many times questions like these must be answered with a court decision.

Classification of social impacts

The social impacts of TWM are any impacts of water management on people other than those impacts that are mainly economic and environmental. For example, if a water project hurts poor people living in a flood zone, that would be a social impact. The economic impact might be small, and sometimes the line between economic and social impacts is hard to draw.

Compared to economics and environmental impacts, social impacts have received relatively little attention. "Water and Related Land Resources: Principles and Standards for Planning," or P&S,[1] initially listed social well-being as an effect, and they listed five categories: real income distribution; life, health, and safety; education, culture, and recreation; emergency preparedness; and other social effects (US Water Resources Council, 1973).

Social rights related to water meet basic human needs

These open-ended categories did not survive to the final versions of the P&S, but they at least offer a starting point to identify categories of social well-being. Using this classification enables us in this chapter to discuss social effects along the lines of the TWM definition. The heading "Public health and safety" covers all public health, safety, and security issues. "Equal opportunity" will cover income distribution, social equity and justice, and access to basic services. (It is treated as an economic impact in chapter 6.) "Community goodwill" will include the following topics from the P&S: life, education, culture, recreation, and other social effects.

This list fits at the base of the "hierarchy of needs" popularized by Abraham Maslow (1943). His hierarchy has physiological needs at the bottom and psychological needs at the top. The concept is that the higher needs apply only after all lower-level needs are met. As shown by Figure 8-2, water needs fit at the lower levels, which meet life needs to breathe, eat, dispose of body wastes, and regulate the body's systems. Health issues such as safe drinking water and clean swimming water fit here too. They are shown at level 2 because of the assignment of health issues to that level, but that does not imply those needs are not essential. Social needs come at higher levels, as shown.

Security needs are next and include public health and freedom from risk of harm, such as from bioterrorism or flood risk. Social needs are higher on the hierarchy, to include convenience issues such as urban

---

[1] See chapter 4.

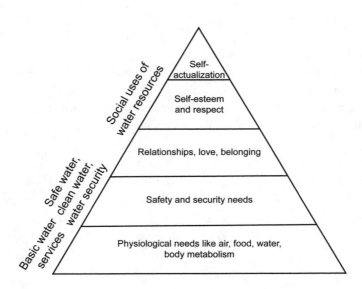

Figure 8-2. Water and the hierarchy of human needs

drainage and water delivered to home and business, as well as the "good life" with water-based recreation, enough water for flowers and lawn, etc. This list of social effects is not exhaustive, but it should include enough to illustrate TWM's role in improving the social effects of water decisions.

A system for social indicators of water projects

The Water Resources Planning Act of 1965 stimulated thinking about how to plan for the social impacts of water projects, as well as for economic and environmental impacts. While neither the act nor its programs have much effect today, some of that thinking is still valid. That is particularly the case for a system developed in the West for social goals and indicators (Technical Committee, 1974).

The committee reviewed potential social goals for water resources projects and developed a system of measurement with nine categories. Each of these can be divided into multiple subgoals, and each subgoal can be measured by several indicators. The report illustrates a total measurement system with five levels, with the indicators increasing by an exponential factor of about three at each level. The result is a measurement system with hundreds of indicators, similar to a regional economic study.

While the detail is too complex for this discussion, a listing of the social goals of water resources development helps us to see how this system fits with the one used in TWM and to organize this chapter (Technical Committee, 1974):

- Collective security
- Environmental security
- Individual security
- Economic opportunity
- Cultural and community opportunity
- Aesthetic opportunity
- Recreational opportunity
- Individual freedom and variety
- Educational opportunity

The social contract in America

The underlying concept for considering social values in water management is the social contract. A *social contract* is an implied agreement that binds people together in a social order so that a nation can be formed. People give up some of their rights to government so that social order for all can be maintained. The idea of a social contract derives from the theories of political philosophers such as Thomas Hobbes, John Locke, and Jean-Jacques Rousseau, the latter who saw it as necessary for democracy to work (Wikipedia, 2007). Once these rights are given up to government, people look at least partly to government to meet some of their basic needs. The extent to which government meets needs is a political decision that involves choices of how many social benefits it should provide.

Meeting basic water needs is part of the social contract in the United States

When considering social effects of water management, we might get into controversies because whether these are provided is not at issue; the issues are how many benefits to provide and who should pay the cost. These are worked out under our social contract through our political and social institutions. For example, in our nation we work together to provide jobs, government provides at least a partial safety net for the basics of life, and people agree to get along with each other or work things out through our system of justice.

Mechanisms to advance society

Social mechanisms stem from our ideas of what is required to advance society. This requires a shared understanding in our nation of the elements of life that work best for everyone. TWM aligns with this thinking in several ways, as seen by phrases in its definition whereby TWM is the exercise of stewardship for the greatest good of society, addresses social values, and fosters community goodwill.

Some social mechanisms that knit our society together and how TWM relates to them are given in Table 8-1. From the list in Table 8-1, we see that many social needs for water arise from our dependence on it to make a good society. We can summarize these needs in the categories of public health and safety, equal opportunity, and community goodwill.

Public health and safety

Public health is a very important goal of water management. This is why TWM's role to foster it is in its definition. Health issues are addressed by the Safe Drinking Water Act (public water supply) and the Clean Water Act (water in the environment). One regulates purity of drinking water and the other regulates discharge of pollutants to waters and protection for fish and wildlife, as well as the recreational use of water.

Water is essential for the body's systems. Without adequate safe drinking water, health impacts and even death result. Water quality is directly involved in preventing disease and providing minerals for the body's functions and bone structure.

> **Water quality is directly involved in preventing disease and providing minerals**

Knowledge of the links between water and health expanded greatly during the nineteenth and twentieth centuries. Public health had been abysmal up through the Middle Ages and even into the Industrial Revolution. Water and health links remained hidden until the development of microbiology. When Dr. John Snow discovered in 1854 that a cholera outbreak in London could be traced to a single source of water, it initiated a revolution in water management. With the new discoveries, filtration started in 1887 and chlorination in 1909. By 1900, waterborne infectious diseases were on the decline, but chemical problems increased, leading to the 1974 Safe Drinking Water Act (Grigg, 2005).

In recent years, Hollywood has produced movies that tell true stories of chemical contamination of water. *Erin Brockovich* was about a legal clerk who helped win a 1993 case against the Pacific Gas and Electric Company of California. The case alleged contamination of drinking water with hexavalent chromium in the southern California town of Hinkley. *A Civil Action* was a 1996 movie (based on a book by the same name) about the contamination of groundwater in New England by trichloroethylene, an industrial solvent.

Even with modern public health systems, water-related threats come when people are exposed to contaminants from eating and drinking, swimming, and inhaling. Healthy people and people with weak immune systems respond differently to threats. Youth are in a growth phase, while the adult population is in various stages of aging and some are more

Table 8-1. TWM contributions to social systems

Kinds of social systems	How TWM contributes to social systems
Hard systems* (infrastructure and natural resources)	Water systems are important parts of the critical infrastructure systems and of natural systems, the basis for sustainable environment
Soft systems (rule of law, organization, government, nongovernmental organizations, institutions)	Water institutions and institutional arrangements are central to governmental functioning and to the vigor of the nongovernmental sector
Justice, compassion, and security (safety nets so that members of society are cared for and anxiety is reduced)	Public health and access to basic water and sanitation systems should be basic human rights, along with security against emergencies
Economic opportunity (robust economy that offers opportunity)	Water inputs are essential to economic advancement
Leadership and inspiration	Leadership in solving water and related problems and environmental education will inspire all ages
Wealth and capital	Water systems and infrastructure comprise important inputs on which to build wealth
Experience (common heritage to rely on for decisions and to solve problems)	National experience in solving complex water problems will create lasting bonds in society
Societal values and ethics	Environmental and social values of water instill unity of purpose in caring for people, resources, and property
Sense of national purpose and progress (feeling that life is significant and meaningful)	Effectiveness in solving water problems, rather than gridlock, will instill a sense of purpose
Finer things of life	Water is used for recreation, art, literature, and the finer things of life

* I am indebted to Audrey O. Faulkner and Maurice L. Albertson (1986) for the terms "hard" and "soft" systems to refer to village development.

susceptible to waterborne threats at these stages. Wealthy people are more protected from threats, while the poor are more exposed to environmental hazards. Consequences of water contamination vary according to severity of impacts, from mild illness to death. Threats to public health focus on outbreaks, although toxicological agents from spills and accidents also affect health.

All sources of drinking water contain some naturally occurring contaminants. However, at low levels, these contaminants generally are not harmful in drinking water. Removing all contaminants would be expensive and not necessarily improve health.

Sudden disease outbreaks can be caused by water contaminants, and some contaminants can cause longer-term chronic conditions, as in cancer. Adverse health effects of water contaminants range across a wide variety of diseases. For example, Table 8-2 illustrates a few of the major effects listed in USEPA's (2003) table of primary drinking water standards (also see Figure 8-3).

> **Social needs for water can be summarized in three categories: public health and safety, equal opportunity, and community goodwill**

Localized outbreaks of waterborne disease have been linked to contamination by bacteria or viruses, probably from human or animal waste. For example, *Cryptosporidium* may pass through water treatment filtration and disinfection processes to cause gastrointestinal disease, sometimes with deadly consequences among sensitive members of the population. Disasters, such as floods, hurricanes, or earthquakes, can also cause waterborne disease outbreaks. Craun, Calderon, and Frost (1998) reported on 35 outbreaks. The 1993 Milwaukee outbreak of *Cryptosporidium* caused illness in 400,000 and a number of deaths (Fox and Lytle, 1996). In 2000, an *E. coli* contamination incident in Walkerton, Ontario, led to seven deaths, and more than 2,000 fell ill—half the town. Floodwaters sweeping over cattle grazing lands and allegations of utility problems were said to be the cause, along with lack of training (AWWA, 2000).

Water supply and sanitation in developing countries remain critical public health issues. Billions of the world's population lack access to safe drinking water and/or sanitation, and they suffer terrible rates of disease and infant mortality. Sustainable development seems like an empty goal without meeting the needs of these people.

The United Nations designated the 1980s as the International Drinking Water Supply and Sanitation Decade with a goal to supply the world's population with safe drinking water and sanitation by 1990. Although the goal was not met and many people still lack basic water services, progress was made. The root causes of the problem are disorder, lack of opportunity, injustice, and poverty. These cause high population growth, lack of opportunity in rural areas, and rural-to-urban migration, leading to shantytowns that go by many different names: squatter areas, slums, informal settlements, illegal settlements, *barrios marginales* (Honduras), *tugurios* (El Salvador), *favelas* (Brazil), *pueblos jovenes* (Peru), *asentamientos populares* (Ecuador), *villas miserias* (Argentina), *bustees* (India), and *bidonvilles* (France or North Africa) (Grigg, 1996).

Daniel A. Okun (1991) reported that in his 50 years of work on the water and sanitation problems in urban areas, they had gotten worse. The reasons were an inadequate supply of water in the cities attributable to

Table 8-2. USEPA list of contaminants and their potential effects

Contaminant	Potential effect
Cryptosporidium, Giardia, and enteric viruses from human and animal fecal waste	Gastrointestinal illness (e.g., diarrhea, vomiting, cramps)
Trihalomethane as a disinfection by-product	Liver, kidney, or central nervous system problems; increased risk of cancer
Arsenic from erosion of natural deposits; runoff from orchards, runoff from glass and electronics production wastes	Skin damage or problems with circulatory systems; possible increased risk of cancer
Copper from corrosion of household plumbing systems; erosion of natural deposits	Gastrointestinal distress from short-term exposure, liver or kidney damage from long-term exposure
Lead from corrosion of household plumbing systems; erosion of natural deposits	Adults may have kidney problems; high blood pressure. Infants and children may have delays in physical or mental development; children could show slight deficits in attention span and learning abilities
Nitrate or nitrite as runoff from fertilizer use; leaching from septic tanks, sewage; erosion of natural deposits	Babies may get blue-baby syndrome
Benzene as discharge from factories; leaching from gas storage tanks and landfills	Anemia; decrease in blood platelets; increased risk of cancer
Polychlorinated biphenols (PCBs) as runoff from landfills; discharge of waste chemicals	Skin changes; thymus gland problems; immune deficiencies; reproductive or nervous system difficulties; increased risk of cancer

Source: USEPA, 2003.

limited water resources, and/or poor facilities for treating and distributing the water compounded by an absence of proper sewerage.

Waterborne infectious agents can reach taps, even when the water is safe as it leaves the treatment plants. Wastewaters are discharged into drainage channels and pollute wells and the groundwater table, creating unsanitary conditions. This is aggravated in the fringe areas, where many poor and landless families live.

Robertus Triweko (1992) attributed the problems in urban areas to two causes: low level of service and inability to improve and maintain the continuity of service. He recommended a comprehensive solution that utilized a management system with technological, institutional, and financial subsystems.

The US Agency for International Development (1990) organized a Water and Sanitation for Health (WASH) project, which yielded valuable

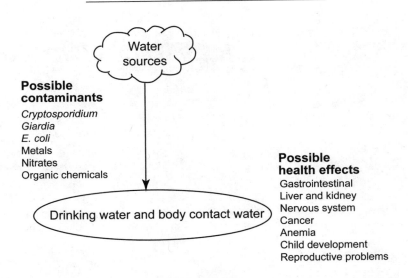

Figure 8-3. Some possible health effects related to water

lessons that align with TWM and the ideas in this book. For example, the measure for success of both a national system for development and the community management systems it creates is sustainability, or the ability to perform effectively and indefinitely after donor assistance has been terminated. This outcome is more likely to occur if each of the key participants recognizes and assumes its appropriate role and shoulders its share of the responsibility. The lessons learned by WASH emphasized institutional issues that are consistent with TWM.

Safety and security

Safety and security goals as social effects of water management include emergency preparedness, security against flood and drought, and protection against exposure to contamination.

Water service providers face growing risks from infrastructure decay, natural disasters, accidents, and malevolent threats. In addition, systems can fail from collateral damage and interdependences with failed systems that cause cascading effects. They have learned many lessons from past disasters and emergencies, but large events, such as Hurricane Katrina or massive power failures, can disrupt entire water systems.

Another safety and security issue is low-income housing in floodplains. Many of the world's population live in vulnerable areas, and they have little choice in selecting safer places to live.

Water utilities must remain vigilant and have comprehensive engineering and management measures to protect against threats, at the same time that they provide reliable and safe water every day.

These comprehensive measures should be embodied in utility-wide risk management programs that are widely understood and include within the programs plans for a range of threats (Grigg, 2006).

Equal opportunity

This category includes income distribution, social equity and justice, and access to basic services. In economic parlance, a fair income distribution depends on equitable distribution of water-related benefits. That is, if investments are made to improve water services, the benefits should flow equitably.

> **Providing water and sanitation is a universal need across the world**

A few brief examples will illustrate. Say a lake is built with taxpayer funds. Wealthy people should not get a lot of benefits by buying lakefront property and lower income people get nothing. In water rate-setting, a principal issue is to set water rates at levels that are fair to all income groups. Another example would relate to irrigation water at a fair price, given the difficulty farmers might have in making an income from crop production.

Social equity and justice is a broad category that includes environmental justice, fair access to water and sanitation for low-income people, fair water rates, and fair allocation of water to all for social uses. As examples, consider that a wastewater treatment plant is to be built in a town. The lowest-cost property might be the logical site, but that might affect low-income people living nearby. Native American water rights are another equity issue that has received a lot of attention in recent decades.

Access to basic services means that, regardless of their ability to pay, all people are entitled to vital water-related services. How this plays out depends on the situation, but it might mean that lifeline rates are used or some other subsidy scheme is created to make sure that people do get the water they need.

Community goodwill

Community goodwill includes education, culture, recreation, and other social effects. It means a sense of harmony and friendliness within a community, which should bring solidarity, contentment, and civic spirit. Since water services are required by communities for a number of purposes, banding together to create desirable ones at reasonable cost offers leaders an opportunity to involve citizens in ways to improve community solidarity and shows the close linkage between water management and public involvement.

> **Social justice is important in water service delivery**

The opposite of community goodwill would be actions leading to disharmony and strife. On this side of the ledger, a bad project that split the community, had cost overruns and high rates, and/or had suspected corruption in a program with other negative outcomes would create distrust and divisions within the community.

Water projects bring with them great opportunities for public education, both on the environmental education side and the science and mathematics side.

Culture, recreation, and other social effects

Water-based recreation can improve community solidarity and enhance life for all citizens. The development of waterways can promote cultural attractions, such as San Antonio's popular River Walk. Water delivered for convenient use in homes and businesses adds to the quality of life, and water used for amenities such as gardens, flowers, and lawns adds to urban life as well. Storm drainage can be used for parks development in high- and low-income neighborhoods. Unexpected recreational benefits such as using public water supplies for recreation during hot days can liven up a community as well.

For the remaining category, other social effects, we can consider that anything that improves people's lives is a social effect. These may be site-specific and go beyond the examples given above.

Social impact analysis: an assessment tool

For water actions, social impact assessment involves preparing inventories of social effects like those explained earlier. To explain social effects in a TBL report, the formal tool of *social impact assessment* (SIA) could be used. It goes along with environmental impact assessment as the main methodology for its category of effects.

SIA is a methodology to review the social effects of projects and other interventions such as new program initiatives.

Just as there are principles for environmental impact assessment, SIA has guidelines, too. These include analysis, monitoring, and managing social consequences of policies, programs, plans, and projects. In a practical sense, you can prepare one by inventorying the categories of social effects (public health and safety, equal opportunity, and community goodwill) and preparing a multicriteria decision analysis matrix to assess how each impact changes under a different water action (Barrow, 2000; International Association of Impact Assessment, 1996; Wikipedia, 2006).

Social rights and social responsibilities

TWM carries with it both social rights and social responsibilities. Social impacts of water actions focus on the rights and can be assessed in categories included in the TWM definition, which can be summarized as public health and safety, equal opportunity, and community goodwill. These are consistent with earlier comprehensive thinking about the social aspects of water development.

> Social impact assessment is another TWM tool for water resources managers

As we see, water management involves a variety of social effects. The way to consider them in TWM is to look at any water action, whether a new project, change in policy, rate structure, or any other change, and consider how it will affect the social variables represented in the categories of public health and safety, equal opportunity, and community goodwill. These will, of course, vary from place to place.

The other side of the coin, individual and corporate social responsibility, is also a core issue in TWM and is explained in chapter 10.

Summary points

- The main reason why water cannot be managed just by economics and the market is that social and environmental water needs are not valued well by market choices.
- TWM addresses social requirements for water in general statements such as that it addresses social values and fosters public health, safety, and community goodwill. However, it is left to the water utility to determine how to accomplish these ideals.
- Although access to water is a basic right, social aspects of water management can be controversial. The responsibility of the water industry to engage society in its water management actions supports important national goals such as lack of discrimination, equal opportunity, justice, basic health care, and access to basic services.
- The social impacts of TWM are any impacts of water management on people other than those that are mainly economic and environmental. They were identified by the US Water Resources Council in categories of real income distribution; life, health, and safety; education, culture, and recreation; emergency preparedness; and other social effects.
- The underlying concept for considering social values in water management is the social contract, which is an implied agreement that binds people together in a social order. Under this social contract, the extent to which government meets needs such as water needs is a political

decision that involves choices of how many social benefits to provide.
- Public health is a very important goal of water management and TWM. The Safe Drinking Water Act regulates the purity of drinking water, and the Clean Water Act regulates the discharge of pollutants to waters and protection for fish and wildlife as well as recreational use of water.
- Knowledge of the links between water and health expanded greatly during the nineteenth century, and by 1900, waterborne infectious diseases were on the decline. However, chemical problems increased, leading to the 1974 Safe Drinking Water Act.
- Safe water supply and sanitation in developing countries remain a critical public health issue. Billions lack access to safe drinking water and/or sanitation and suffer high rates of disease and infant mortality.
- Safety and security goals as social effects of water management include emergency preparedness, security against flood and drought, and protection against exposure to contamination. Another safety and security issue is low-income housing in floodplains. Many of the world's population live in vulnerable areas, and they have little choice in selecting safer places to live.
- Equal opportunity as a social value of water management includes income distribution, social equity and justice, and access to basic services. Social equity and justice is a broad category that includes environmental justice, fair access to water and sanitation for low-income people, fair water rates, and fair allocation of water to all for social uses. Access to basic water-related services is important to public health and social justice.
- Social impact assessment, a methodology to review the social effects of projects and programs, can be used to explain social effects in a TBL report.

Review questions

1. If water cannot be managed just by economics and the market, how are social and environmental water needs taken care of?

2. It is fine to say that access to water is a basic right, but how does society pay for it?

3. Under the social contract in the United States, the extent to which government pays for water projects and programs is a political decision. Is it fair to use federal funds, which are mainly derived from income taxes, to pay for projects in different parts of the country?

4. How does the Clean Water Act affect the quality and safety of drinking water?

5. Compared to the nineteenth century, what have been the major drinking water quality issues of the twentieth century?

6. Describe the urgency and importance of safe water supply and sanitation in developing countries.

7. Explain how you would use social impact assessment in a TBL report.

References

American Water Works Association. 2000. Walkerton Coroner's Office Investigating Deaths. *Mainstream* 44 no. 7: 1, 3.

Barrow, C.J. 2000. *Social Impact Assessment: An Introduction*. London: Arnold.

Bok, Derek. 1996. *The State of the Nation: Government and the Quest for a Better Society*. Cambridge, Mass.: Harvard University Press.

Craun, Gunther F., Rebecca L. Calderon, and Floyd J. Frost. 1998. An Introduction to Epidemiology. *Jour. AWWA* 88 no. 9 (September): 54–65.

Faulkner, A.O., and M.L. Albertson. 1986. Tandem use of Hard and Soft Technology: an Evolving Model for Third World Village Development. *International Journal of Applied Engineering Education* 2 no. 2: 127–137.

Fox, Kim R., and Darren A. Lytle. 1996. Milwaukee's Crypto Outbreak: Investigation and Recommendations. *Jour. AWWA* 88 no. 9 (September): 87–94.

Grigg, N. 1996. *Water Resources Management: Principles, Regulations, and Cases*. New York: McGraw-Hill.

Grigg, N. 2005. *Water Manager's Handbook*. Fort Collins, Colo.: Aquamedia Publications.

Grigg, N. 2006. Disaster Preparedness and Emergency Response in Water Utilities. Special edition, *Jour. AWWA* 98(3) 242–246, 249–255.

International Association of Impact Assessment. 2006. *Principles of Impact Assessment*. http://www.iaia.org. Accessed November 15, 2006.

Maslow, A.H. 1943. A Theory of Human Motivation. *Psychological Review* 50: 370–396.

Okun, Daniel A. 1991. Meeting the Need for Water and Sanitation for Urban Populations. The Abel Wolman Distinguished Lecture, National Research Council, May, Washington, D.C.

Technical Committee of the Water Resources Research Centers of the Thirteen Western States. 1974. *Water Resources Planning, Social Goals, and Indicators: Methodological Development and Empirical Test*. Logan, Utah: Utah Water Research Laboratory.

Triweko, Robertus. 1992. A Paradigm of Water Supply Management in Urban Areas of Developing Countries. Ph.D. Diss., Colorado State University, Fort Collins.

US Agency for International Development. 1990. *Lessons Learned from the WASH Project*. USAID Water and Sanitation for Health Project. Washington, D.C.: US Agency for International Development.

US Environmental Protection Agency. 2003. *List of Drinking Water Contaminants & MCLs*. www.epa.gov/safewater/mcl.html. Accessed January 15, 2004.

US Water Resources Council. 1973. Water and Related Land Resources: Principles and Standards for Planning. Part III. *Federal Register* 38 no. 174 (September 10): 24778–24869.

Wikipedia. 2006. *Social Impact Assessment.* http://en.wikipedia.org/wiki/Social_Impact_Assessment. Accessed November 15, 2006.

Wikipedia. 2007. *Social Contract.* http://en.wikipedia.org/wiki/Social_contract. Accessed May 23, 2007.

CHAPTER 9

LAWS AND REGULATIONS OF WATER MANAGEMENT

In a perfect world, TWM would work purely on a voluntary basis. However, the reality is that stewardship has its limits. For this reason, regulatory controls and law enforcement are required to make TWM work. That is why I suggested in chapter 3 that a statement be added to the TWM definition to recognize compliance with laws and regulations.

In the United States, a body of mostly federal and state law frames our choices in water management. These have been interpreted through many court decisions, so that water law now comprises a rich body of legal instruments through which water is managed. This chapter presents a digest of the key laws and regulatory controls that govern water management. Emphasis is on how the water laws control water decisions and create a system of regulatory control of water management.

To illustrate how laws and regulations have increased their grip on water management choices, Figure 9-1 shows the growth of water laws.

In the final analysis, laws and regulations can only be partially successful in achieving the goals of TWM. They emphasize the negative side of it and limit what people can do. The other side, to encourage people to do more than the minimum, is outside the realm of laws and more in the realm of incentives. These are explained in the next chapter.

Each type of water decision is subject to a wide array of legal controls. Capital decisions require that projects pass through hoops established by environmental laws. Regulatory decisions determine how much money is required to meet standards and are key elements in cost effectiveness. Operating decisions require compliance with environment, health, and safety, and also affect costs. Financial decisions face a different set of laws and regulations that are passed by all three levels of government.

How water laws work in the United States is reflected in other countries as well. For example, all developed countries will have laws governing ownership of water rights or permits to use water, as well as some version of a safe drinking water law. Another example is how comprehensive issues of water management are embodied in the Water Framework Directive of the European Union.

Law coordinates and regulates water management

Chapters 3 and 4 explained how coordination mechanisms are necessary to achieve balance in water decisions and management. While people sometimes want a "czar" to make the decisions, the reality is that government mechanisms provide coordination for decisions, and often the mechanism is law, regulations, or the judicial system. Unfortunately, this often makes adaptable management difficult to achieve.

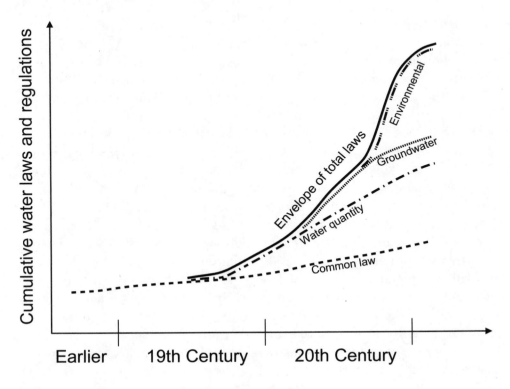

Figure 9-1.   How water-related law has grown

Table 9-1. Responsibility by element of the TWM definition

TWM element	Responsibility
Exercise stewardship for the greatest good of society and the environment	Shared
Require participation of all units of government and stakeholders	Shared
Encourage planning and management on a dynamic basis that adapts to changing conditions and local and regional variations	Shared
Balance competing uses through efficient allocation	Public
Conduct decision-making through process of coordination and conflict resolution	Shared
Promote water conservation, reuse, source protection, and supply development to enhance water quality and quantity	Shared
Address social values, cost-effectiveness, environmental benefits and costs	Shared
Foster public health, safety, community goodwill	Shared

To see why this is necessary in the US system, let's revisit the elements of the TWM definition.[1] By reviewing the TWM definition in Table 9-1, which distinguishes among its elements, we see that every responsibility implied by the definition is either shared among utilities and public bodies or is mostly a public responsibility.

Responsibilities that belong to the public are mostly taken on by government, and this creates a major problem for the stewardship aspect of TWM. Ensuring that most of its elements occur is a public or common responsibility, and only law and regulations address this, not any form of unified or common action.

Relying so heavily on regulations is like putting a system on a rigid autopilot system with limited if any means to adapt to changing conditions. This can create a barrier to the element of TWM that seeks a "dynamic process that adapts to changing conditions and local and regional variations." The reality is, however, that laws and regulations go beyond just controlling actions and can provide management instruments to aid planning and coordination. For example, in obtaining new water supplies, Graham et al. (1999) included state protection of instream flows, coordination of surface water and groundwater, protection of endangered species, growth management, Clean Water Act (CWA), Wild and Scenic Rivers, interbasin transfer (IBT), tribal and reserved water rights, treaties, wetlands law, and navigation rights.

Another problem with the regulatory system is that it might work well with large players, who can afford substantial investments, but many

---

1 See chapter 3.

of the actions that affect sustainability are carried out by small players or individuals.

How law determines management choices

Law and regulations are used during each phase of water management: planning, implementation, and operation of facilities. During the planning phase, law and regulations set limits on options that can be considered for water facilities. In implementation, laws and regulations governing finance, design, and construction take effect. In operations, laws and regulations exert control over each decision that occurs.

To explain how laws affect the range of choices for water management, Figure 9-2 illustrates how water laws align with the processes of water management and influence decisions by water managers. The figure shows examples of specific points in the water management chain where principal laws and regulations exert control, or, in other words, laws coordinate water management actions. Starting from the left of the diagram, you can see that reservoir authorizations and rules on interbasin transfer (IBT) take effect. The Clean Water Act's Section 404 regulations affect wetlands in a tributary area. Reservoir releases are subject to laws such as for instream flows (ISF) and the Federal Power Act (FPA), and state law govern diversions from the stream. Also, seven-day, ten-year low flows (7Q10) and Endangered Species Act (ESA) rules govern streamflow for water quality and environmental purposes. The local floodplain ordinance governs land use in the floodplain.

The Safe Drinking Water Act (SDWA) will govern water treatment, and water use restrictions will govern operation of the distribution system. Industrial pretreatment rules, as well as local ordinances, govern discharges to the collection system. The wastewater treatment plant (WWTP) complies with the National Pollutant Discharge Elimination System (NPDES) permit program of the Clean Water Act (CWA).

Industries and farms that pump from wells must comply with state or local groundwater restrictions. The farm and the city must also comply with CWA rules on irrigation and stormwater return flows. City B has a private water company that is also regulated by the state's public utilities commission (PUC).

These acronyms make an alphabet soup, but their occurrence is a reality of the highly regulated and government-dominated world of water management.

CHAPTER 9   LAWS AND REGULATIONS   211

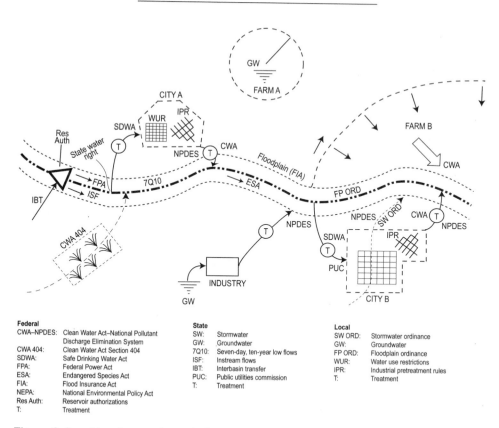

Figure 9-2.   How law and regulations affect water management

Laws versus regulations

In general, law refers to all law, of all levels and types, that society requires to function. The authority for law can be legislative, judicial, or executive. Law can be based on statutes or common law, meaning it is based on accepted standards of doing things. Operating under the rule of law means that society is controlled by laws in an orderly way. Without the rule of law, chaos will reign. So *law* means the whole collection of rules that govern how society functions.

Regulations are rules or directives that are issued by administrative agencies and backed by the authority of law. An example would be the Surface Water Treatment Rule, developed by USEPA to implement part of the Safe Drinking Water Act. Regulations are important but have less authority than law because they can often be challenged at lower levels or are subject to more flexible interpretation.

The water industry is regulated and must comply with rules for:
- Health and safety (such as to supply safe drinking water)

- Water quality (such as to maintain clean streams)
- Fish and wildlife protection (such as to provide instream flows for fish)
- Quantity allocation (such as to recognize legal water rights)
- Finance (such as to control rates of a private water company)
- Service quality (such as to maintain an adequate water pressure)

Water law

Water law is all law relating in any way to water. Since water touches many sectors, water law cuts across types of law, but it mainly means law for control of water and its effects on the environment. Table 9-2 shows examples of how water law is found in several categories of law:

Table 9-2. Examples of legal categories including water law

Type of water-related law	Example of legal/regulatory tool
Water allocation	Permit for water use
Environmental control	Instream flow law
Public health	Safe drinking water law
Energy	Hydroelectric power license
Land use	Drainage control law

Law is made by different levels of government and appears in different forms. Table 9-3 shows a matrix of how law is in constitutions, statutes, administrative rules, and court decisions at all three levels of government. When water is involved, all of these can be considered as water laws.

Table 9-3. The legal matrix by level

Level	Constitutions	Statutes	Regulations	Case law
Federal	Federal Constitution	Federal law	Federal regulations	Federal cases
State	State Constitutions	State law	State regulations	State cases
Local	Local Charters	City codes	Local regulations	Local cases

For example, in Colorado, the state Constitution establishes the state's system of using the appropriation doctrine to allocate water rights. Federal statutes establish most environmental laws, and the Supreme Court and other courts set decrees that determine how some water systems must operate.

Water law classification

Law books present statutes and cases of water law in categories such as these:[2]

- Surface water allocation and use law
- Groundwater law
- Diffused surface water and drainage law
- Federal and Indian Reserved Rights
- Navigation law
- Water power law
- Federal agency laws
- Environmental laws
- Health laws
- Laws governing public organizations

However, these categories may not be a useful classification system for water managers, who need to know how the law affects their management decisions and prerogatives. More useful would be a framework more oriented toward management tasks. Table 9-4 shows such a framework.

Table 9-4. Legal frameworks for water management tasks*

Management task	Legal framework
National and basin coordination	Law specifying national coordination, policy, standards, and assessments. Laws relating to data. Basin and large rivers planning and coordination. Transboundary issues. Drought response. Dispute resolution. Intergovernmental coordination.
Water use	Water supply access and water rights ownership. Surface water and groundwater. Security of title. How rights are transferred.
Safe drinking water	Public health protection.
Clean water in environment	Regulation of stream and groundwater quality.
Public services	Regulatory control of public service delivery.
Finance	Laws enabling financial assistance to water authorities.
Habitat preservation	Environmental and instream flow law.
Coastal water management	Law to coordinate land use and water management in coastal areas.
Flood and security	Flood control, stormwater, security, and emergency management.

Source: Grigg et al., 2004.

*For the ideas behind this framework, the author thanks his team of engineers, lawyers, and economists on the project reported in Grigg et al., 2004.

---

[2] As examples, see Getches, 1990, and Corbridge and Rice, 1999. These are excellent texts, but they are designed for lawyers and law students more than for water managers. Rice and White (1987) is an example of a text written for both managers and lawyers, albeit at an operational level.

Water laws by levels of government

The nation's history explains why some laws are state and some are federal. States rights were dominant when the US Constitution was drafted, and federal–state relations have been a contentious issue ever since. Because of this legacy, property rights in water or permits to use it are under state law. Environmental law, which came along later, is mostly federal law. In this section the main water laws are listed by the level of government at which the authority resides.

Federal laws

The principal federal laws that affect water management deal with interstate and international issues, broadly defined. The need to integrate areas of policy such as environmental and interstate commerce gives the federal government a license to become involved in what some consider state and local affairs. The federal laws explained in this section are the Clean Water Act, Safe Drinking Water Act, National Environmental Policy Act, Endangered Species Act, Federal Power Act, Flood Insurance Act, and Authorizations and Appropriations for Federal Water Projects.[3]

Clean Water Act

If each state had its own Clean Water Act (CWA) and there were no federal law, pollution around the country would be very uneven. Some states would have low standards and it would be a divisive situation. That is why one of the goals in the federal CWA is to equalize conditions among states so that one state could not attract industry from another by offering more lenient environmental laws.

The CWA is the main US law governing environmental water quality. It regulates discharges of pollutants into the waters of the United States, and it also governs related effects, such as protection of wetlands and regulation of dredging in streams. This law has had dramatic effects on water management, on industry, on cities, and on the environment. The goal of the CWA is consistent with TWM: "to restore and maintain the chemical, physical, and biological integrity of the nation's waters so that they can support the protection and propagation of fish, shellfish, and wildlife and recreation in and on the water."

When originally passed in 1972, the CWA was called the Federal Water Pollution Control Act Amendments. After being amended in 1977,

---

3 There are more federal water laws, for example the Wild and Scenic Rivers Act. For more complete explanations see textbooks, such as Getches, 1990 or Goldfarb, 1988. Also, Web sites of federal agencies, such as EPA, have summaries of the laws.

it became known as the Clean Water Act (US Environmental Protection Agency, 2003 a, b). The CWA uses regulatory and nonregulatory tools to reduce pollutant discharges into waterways, to finance municipal wastewater treatment facilities, and to manage polluted runoff. It does not deal directly with groundwater, which is mostly left to the states.

Early CWA efforts focused on the chemical pollution and on point sources. Now, more attention is given to physical and biological integrity and to nonpoint sources. CWA programs have also shifted toward watershed-based strategies, meaning that more holistic approaches are taken. Again, it is supportive of TWM methods. Each program element of the CWA involves a large effort, as shown in Table 9-5.

The Clean Water Act is the main law governing environmental water

Safe Drinking Water Act

The Safe Drinking Water Act (SDWA) applies health-based standards to protect drinking water against threats from improper disposal of chemicals, animal wastes, pesticides, human wastes, wastes injected underground, and natural substances. Much of the material in this section is summarized from USEPA guides about the SDWA (USEPA, 1999); AWWA publishes the comprehensive *Field Guide to SDWA Regulations*, and the *Journal AWWA* regularly publishes explanatory columns regarding different aspects of the act.

The SDWA was originally passed in 1974 as the successor to earlier regulatory programs of the US Public Health Service and was amended in 1986 and 1996. It provides a framework for USEPA, states, tribes, water systems, and the public to work together to provide safe water. Originally, the SDWA focused primarily on treatment, but the 1996 Amendments added source water protection, operator training, funding, and public information.

Like the CWA, the SDWA has had widespread and dramatic effects on cities and on the water management industry. As explained in chapter 3, the SDWA applies to all public water systems in the United States that provide piped water for human consumption and have at least 15 connections or regularly serve at least 25 people. There are about 170,000 of these public water systems, including about 55,000 community water systems (CWS) and a larger number of noncommunity water systems (NCWS). CWS are mainly homes, apartments, and condominiums and serve the same people year-round. NCWS are for transient facilities (such as rest areas and campgrounds) and nontransient facilities (such as schools with their own water systems).

Table 9-5. Programs of the Clean Water Act

CWA program	How it works
Antidegradation policy	If all standards are met, antidegradation policies keep water quality at acceptable levels.
Assessments	In addition to riverine waters, the quality of reservoirs, lakes, estuaries, and coastal waters is assessed.
Authority	USEPA has authority to implement water pollution control programs. The 1977 Amendments encouraged states to take control of the permit system.
Changes	Other laws have changed parts of the Clean Water Act. For example, the Great Lakes Critical Programs Act of 1990 implemented the Great Lakes Water Quality Agreement of 1978.
Dredge and fill regulations (Section 404 program)	Section 404 regulates the placement of dredged or fill materials into wetlands and other waters of the United States. It delegated authority to the US Army Corps of Engineers to administer the program.
Effluent standards	Effluent standards regulate levels of contaminants for discharge. They are set after analysis of stream capacity to assimilate wastes and to prohibit other harmful discharges. They become part of the permit conditions for dischargers.
Enforcement	The act provides authority to enforce its provisions.
Financing and construction of sewage treatment plants	The act authorized a construction grants program for wastewater treatment plants. The 1981 amendments reduced federal financial support and the 1987 Water Quality Act replaced the grants with Clean Water State Revolving Funds.
Monitoring and assessment	Water bodies are monitored to determine if standards are met.
Nonpoint source pollution	Section 319 addresses nonpoint sources, largely through grants. The 1981 act required USEPA to develop regulations for stormwater control and for states to prepare nonpoint source management programs.
Permit program for point sources	The National Pollutant Discharge Elimination System (NPDES) permit program covers point sources of pollution discharging into surface waters.
Planning	The Act created a Section 201 regional planning program and Section 208 water quality planning program.
Priority pollutants	A 1976 consent decree required USEPA to publish criteria for specified pollutants by 1979. These were later designated by Congress as toxic pollutants under Section 307(a) of the act. By 1994, 99 of the 126 USEPA-selected key chemicals or classes of chemicals had been designated. USEPA continued to work on criteria for sediment, organisms that feed on material in the sediment, aquatic life that feeds on the organisms, and the health of humans who ingest the aquatic life (USGAO, 1994).
Reports on condition of the nation's waters	The 305(b) Report is issued every two years to report on healthy, threatened, and impaired waters. The 303(d) List includes waters that are either threatened or impaired.

Table 9-5. Programs of the Clean Water Act

CWA program	How it works
Total maximum daily loads (TMDLs)	If a water body is not meeting standards, a strategy is required. The TMDL concept was developed for this purpose. TMDLs determine the load that is consistent with standards and enables allocation of loads among sources.
Water quality certification	Section 401 requires federal agencies to obtain certifications that discharges will not result in violations of water quality standards.
Stream water quality standards	The act establishes water quality standards for streams.

Oversight of the SDWA is mainly through state drinking water programs, and states can receive *primacy* to operate the SDWA program. This involves testing, reviewing plans for improvements, inspections and surveys, training and technical assistance, and enforcement actions.

The SDWA applies multiple barriers to protect water: source water protection, treatment, distribution system integrity, and public information. If the water does not meet standards, the water system operator is obligated to inform the public. Water system operators are also required to issue annual consumer confidence reports. These report the source of the water, detected contaminants, and possible health effects. The SDWA also provides for watershed and wellhead protection and for the Underground Injection Control program, which controls the injection of wastes into groundwater.

Standards are set by USEPA, which uses science to assess risk to health of the contaminants. It especially considers the most sensitive populations, such as infants, children, pregnant women, the elderly, and people with weak immune systems. USEPA is currently assessing risk for microbial contaminants (such as *Cryptosporidium*), by-products of disinfection, radon, arsenic, and groundwater without disinfection.

Standards are classified as *primary* and *secondary* standards. Primary standards govern contaminants that threaten health, and secondary standards address those that threaten "welfare." The National Primary Drinking Water Regulations (NPDWRs) include mandatory levels (called maximum contaminant levels, or MCLs) and nonenforceable health goals (called maximum contaminant level goals, or MCLGs) for each included contaminant.

The focus of the Safe Drinking Water Act is public health

Under the 1974 SDWA, USEPA was to adopt revised NPDWRs by 1977. However, it took longer. By 1990, USEPA had issued regulations for microbiological contaminants, radionuclides, volatile organic chemicals, fluoride, surface water treatment, synthetic organic and inorganic chemicals, lead, and copper. Secondary drinking water standards were set in 1979. Later, monitoring requirements for corrosion and sodium were set, and rules for trihalomethanes (THMs) were developed.

Highlights of the 1986 Amendments to the SDWA include the requirement to set MCLs for 83 contaminants, to set MCLs for a list of priority pollutants updated every three years, to establish criteria for filtration of surface waters, and to require disinfection for all public water supplies.

Features of the 1996 Amendments were requirements for consumer confidence reports, benefit–cost analysis, the drinking water state revolving fund, more protection against microbial contaminants and disinfection by-products (DBPs), operator certification, public information and consultation, help for small water systems, and source water protection programs.

Under the SDWA, secondary standards are set for contaminants that are not health-threatening. Problems might cause undesirable tastes or odors, have cosmetic effects, or damage water equipment. Public water systems test for them on a voluntary basis.

Odor and taste are used to indicate water quality, but methods to measure them are subjective. Color may indicate dissolved organic material, inadequate treatment, high disinfectant demand, and the potential to produce excess disinfectant by-products. Foaming may be caused by detergents. Skin discoloration may result from silver ingestion, and tooth discoloration may be from excess fluoride exposure. Corrosion can affect the aesthetic quality of water and have economic implications from corrosion of iron and copper and staining of fixtures. Corrosion of pipes can reduce water flow. Scale and sedimentation can also have economic impacts. Scale deposits that build up in hot water pipes, boilers, and heat exchangers restrict water flow. Loose deposits in the distribution system or home plumbing cause sediments to occur (USEPA, 2003c).

Drinking water standards are also used to develop regulations under other statutes, such as the Resource Conservation and Recovery Act, or RCRA. This affects cleanup of contaminated sites and storage and disposal of waste materials.

Under the SDWA, the states and USEPA prepare annual summary reports of compliance with safety standards. The SDWA will continue to have strong and permanent effects on the US drinking water industry. It drives the industry's regulatory structure, finance, research, and product development.

National Environmental Policy Act

The National Environmental Policy Act (NEPA), passed in 1970, establishes goals for environmental policy and the requirement for environmental impact statements (EIS) for major federal actions that affect the environment. NEPA also established the Council on Environmental Quality (CEQ) to review policies and programs for conformity with NEPA and to prepare the president's annual environmental report to Congress.

An EIS evaluates the environmental impacts of a proposed action, unavoidable adverse environmental effects, and alternatives available to the proposed action. In preparing an EIS, the agency consults with other federal agencies with expertise on environmental impact. The president's environmental report describes the condition of the nation's air, aquatic, and terrestrial environments and the effects of these environments on the social, economic, and other requirements of the nation.

NEPA is a powerful instrument for TWM

Since NEPA was passed in 1970, the EIS process has influenced many projects and actions. On the positive side, it provides for coordination of the inputs of diverse interests and thus improves planning. On the negative side, the process can be bureaucratic, expensive, and time-consuming.

Endangered Species Act

The Endangered Species Act (ESA) was passed in 1973 and amended several times. Arguably, it has had the greatest effect on water management of any of the environmental laws. In establishing the ESA, Congress found that some species have been extinguished and others threatened with extinction. When species are listed as threatened or endangered, recovery plans are required to protect habitat for these species of fish, wildlife, and plants.

Using scientific or commercial data, the secretary of the Department of the Interior (USDI) is required to list species as endangered or threatened from habitat destruction, overutilization, disease or predation, inadequacy of regulatory mechanisms, or other natural or man-made factors. Within USDI, the US Fish and Wildlife Service (FWS) implements the act for all species except ocean species, which are the responsibility of the National Marine Fisheries Service.

The agencies determine if a species is near extinction, and if so, they list it as "endangered"; those listed as "threatened" receive a slightly lower level of protection. The agency is then required to devise recovery plans for the species. The secretary of the interior develops recovery plans with management actions, measurable criteria, and time and cost estimates.

Section 7 of the ESA provides strong administrative authority to implement the act. It outlines procedures for interagency coordination to conserve federally listed species and designated critical habitat and requires federal agencies to use their powers to promote conservation of listed species. If under the ESA, the FWS issues a *jeopardy opinion*, then a *recovery plan* is necessary, with elements as required by the federal agencies (University of New Mexico School of Law, 2003).

Water managers are naturally reluctant to have Section 7 implemented because it initiates a different control environment and subjects them to more stringent controls on their actions.

Federal Power Act

The Federal Power Act (FPA), dating from 1920, provides for federal regulation and development of water power, and it controls licensing of nonfederal hydroelectric power (hydropower) generation.

The FPA is administered by the Federal Energy Regulatory Commission (FERC), formerly the Federal Power Commission, which has responsibility to license nonfederal hydroelectric power projects and to regulate interstate sale and transmission of power. Licenses are limited to 50 years.

The Federal Power Act contains provisions for comprehensive river basin planning

The FPA requires water planning because projects must be adapted to a comprehensive plan for improving or developing a waterway. In permit actions, FERC is required to consider both how the project is adapted to the plan and the recommendations of relevant federal and state agencies and Indian tribes. It must consider interstate or foreign commerce, water power development, fish and wildlife, and beneficial public uses, including irrigation, flood control, water supply, and recreation.

The FPA requires FERC to give equal consideration to environmental issues. Licenses must contain conditions that adequately and equitably protect, mitigate damages to, and enhance fish and wildlife affected by the development, operation, and management of projects.

Hydropower relicensing will continue to be an important part of water management. In it, the FPA must be coordinated with a number of other federal statutes and with state law. In 1994, the Supreme Court expanded the regulatory authority over hydro projects to state governments. They ruled that under the Clean Water Act states can accomplish goals such as to preserve fish habitat (Barrett, 1994). In that case, the state's department of Ecology had set a minimum streamflow requirement higher than the project design would allow, arguing that the higher flow was necessary to protect fish.

FERC may exempt from the licensing provisions facilities that use only the hydroelectric potential of man-made water transmission and distribution conduits if they are not primarily for generating electricity (University of New Mexico School of Law, 2003).

National Flood Insurance Act and local floodplain ordinances

The nation's flood management programs are controlled under the National Flood Insurance Act of 1968 and the companion Flood Disaster Protection Act of 1973 (FEMA, 2003). As will be explained in chapter 11, this was a new era in how the nation responded to flood risks.

The National Flood Insurance Act made flood insurance available and created the Federal Insurance Administration, which now resides within the Federal Emergency Management Agency (FEMA). The Flood Disaster Protection Act of 1973 made flood insurance mandatory for property located in Special Flood Hazard Areas.

These laws evolved during a long period of increasing federal attention to flood problems. As far back as the 1850s, the federal government was surveying the Mississippi River to determine options to control floods. By 1890, the lower river was divided into state and local levee districts. In 1913, a flood in the Ohio River Valley killed more than 400 people, and a House Committee on Flood Control was created by Congress, leading to the 1917 Flood Control Act. Over the years, the limits of structural concerns with flooding became more apparent. Spurred by Gilbert White's 1942 study, the government began to consider a different approach to flood policy, and by the 1950s, the government was considering a flood insurance program. This led to the 1968 and 1973 acts (FEMA, 2004).

Flood law affects environmental issues through floodplain management

The goals of the Flood Insurance Act are focused on nonstructural solutions, and FEMA operates an array of programs to achieve them, including the flood insurance program. Local floodplain ordinances are required to implement the flood insurance program. These typically regulate land use in the 100-year floodplain.

Authorizations and Appropriations for federal water projects

Water Resources Development Acts (WRDAs) provide the basic authorization for federal projects and other water management provisions added by Congress. The WRDA of 1986 introduced new reforms for project planning and cost sharing. As it relates to Corps of Engineers programs, the WRDA of 1990 created a new interim goal of no overall net loss of the nation's remaining wetland base and a long-term goal of enhancing all of the nation's wetlands. It also directed the secretary of the army to include environmental protection as a primary mission of the corps. The 1996 and 2000 acts included provisions for the comprehensive Everglades Project.

Treaties and interstate compacts

Water flowing across state or national lines requires treaties, or *interstate compacts*, and can lead to *transboundary conflicts*. Problems of interstate or international streams can involve both water quantity and quality. Water quantity has traditionally been the most urgent because of supply issues for cities and agriculture. In the United States, interstate water quality problems are covered by uniform federal stream standards. However, within states the stream standards are set by the respective states, so the potential for problems exists.

Normally, transboundary issues should be handled in the context of river basin management, and when more than one state is involved, the complexity grows. If voluntary agreement breaks down, formal compacts may be required. These are common in the West but also occur in the East as, for example, in the case of the Delaware River involving New York, Pennsylvania, Delaware, and New Jersey. In the West, the most famous compact is for the Colorado River, which involves seven states and Mexico.

State laws

Important state laws that affect water management are: surface- and groundwater allocation, instream flow laws for water quality and environment, interbasin transfer law, and public utility commission law. Some states, such as California, also have laws like NEPA. These are called SEPAs, or State Environmental Policy Acts.

Surface water allocation

Water allocation law has important roles in water management, particularly in the Western United States. In a basic sense, it deals with the right to use a quantity of water. For example, if a city needs a reliable water supply for its population, it must gain a legal right to divert the water when it is available.

While Western states have used this type of law for many years, water allocation law is now becoming more important in humid states because of increasing population and competition for water. In the Western states, administration of water allocation law is normally done through the state engineer's office or some variation of this office. In Eastern states, systems for administering water allocation law are not as well organized as in the West, but they are evolving.

In the United States, the main systems of state water allocation law are riparian, appropriation, and hybrid systems. The Eastern states mainly follow the riparian doctrine. Nine states follow the appropriation doctrine (Alaska, Arizona, Colorado, Idaho, Montana, Nevada, New Mexico, Utah, and Wyoming). Another 10 follow hybrid systems (California, Kansas, Mississippi, Nebraska, North Dakota, Oklahoma, Oregon, South Dakota, Texas, and Washington). Hawaii follows a system based on historical precedents of their former kingdom, and Louisiana still follows a system based on the French Civil Code (Getches, 1990).

The origins of the riparian doctrine are from Europe. Both England and France have versions of it. Its origin in the United States is in common law, meaning that it has generally not been enacted in statutes or state constitutions but is used as the basis for court decisions.

Water allocation law controlled by states affects TWM in big ways

In the *riparian* doctrine a person whose land abuts the water is a riparian landowner. This owner has rights to the flow of the stream and to make a reasonable use of the water body, as long as other riparians are not damaged (Getches, 1990). In the pure riparian doctrine, the natural flow rule would entitle the landowner to the flow of the stream "undiminished in quantity or quality."

The pure riparian doctrine is not practical, and the doctrine has mostly given way to a *reasonable use* doctrine. It allows landowners to use waters if they do not interfere with reasonable uses of other landowners. This evolves into a set of *administrative systems,* which are patchworks of riparian doctrine and practical, politically acceptable methods for allocating water.

Administrative systems usually use permits, which are like water rights in that they entitle the holder to use the water. However, they are not property rights. Permits usually entitle the holder to withdraw a quantity of water, as for example a permit to withdraw water for a city of 50,000. Conditions on such a permit would be negotiated between the administrative agency and the diverter.

These systems may not answer important questions that arise during shortages. These are usually dealt with in drought response plans. Also, the systems are challenged to deal with providing instream flows and managing interbasin transfers. Other difficult questions deal with security for the permit holder.

States in the West generally use the prior *appropriation* doctrine or a variation of it. This doctrine deals with the fact that there usually is not enough water to satisfy all users, so a system of allocation is needed. It found its way into state constitutions and now is the main principle for water allocation in 19 Western states, although 10 of these states follow hybrid approaches.

It is important to note that the doctrine provides that the water belongs to the public, but the doctrine provides for the right to use it, in order of priority, as long as it is being applied to a beneficial use. Public ownership of water is an important principle behind water management, and debates about it occur in the realm of law and in the concepts of economics.

A water right must be initiated and perfected by appropriation and adjudication. A valid appropriation includes the intent to apply the water to a beneficial use, an actual diversion, and a demonstration of application to the beneficial use. Water rights are administered by a system of rules and regulations, including calls on the river. Rice and White (1987) describe how the administrative systems work. Water rights can be lost through forfeiture or abandonment. Systems are necessary to transfer water rights to new owners or new uses.

In administering the appropriation doctrine, numerous practical problems arise. Imagine trying to precisely determine each water right owner's entitlement in a stream that rises and falls according to hydrologic variation, with uncertain routing of flows from one point to another, unknown return flows, variable weather, and everyone diverting and releasing water according to schedules not under the control of the administrators. While the appropriation doctrine has flaws, it seems destined to continue in use in the West, with periodic tune-ups to respond to pressing needs.

The riparian doctrine seems of little effect in the development of systems to allocate water. States that follow it seem to be focusing on permit approaches and are only now grappling with basic questions of law, such as whether rights to use water belong to the people or become private property subject only the marketplace. Legal systems to answer questions of water rights under this doctrine are apt to be settled piecemeal by each state, with the resulting management systems becoming administrative patchworks.

Environmentalists are concerned that unless states develop adequate systems for water control, considering both quantity and quality, the incremental impacts on ecosystems can be severe. Battles for water control often shift to the environmental arena and away from the arena of water allocation law.

Groundwater allocation and quality law

Groundwater law is a hybrid of water allocation law, resource law, and land use law, and it works differently than surface water law. States take different approaches to groundwater quantity law. In some states, little control is exercised over quantities of groundwater pumping. In other states, sensitive areas are brought under control due to public pressure. For example, an area that develops water table problems might be considered for regulation of pumping, as in coastal zones where saltwater intrusion might occur. In other cases, regulation might be imposed where groundwater levels drop so fast that pumpers are threatened, as occurs in some groundwater management districts.

A state might recognize ownership of the water by the owner of the overlying land but also limit use to reasonable levels to recognize the interdependence of adjacent landowners. Local groundwater ordinances may be implemented by county governments or by management districts with authority to limit pumping. This might occur under the authority of state law, in recognition of the need to regulate groundwater use for the common good.

While groundwater is the principal source for drinking water for more than 50 percent of the population, there is no centralized national groundwater policy. During the 1980s, studies debated whether such a policy was needed, but in the end the result was to rely on the different controls already in place. For example, the SDWA's provisions include the Underground Injection Control program. All states monitor potable groundwater supplies, and most states conduct additional monitoring near potential polluters or in vulnerable areas. Some state laws cover permitting and surveillance of potential contamination sources; siting of waste treatment and other source facilities; promotion of proper pesticide and fertilizer use and animal waste management; and prohibition of land disposal of toxic sludge and hazardous waste. Also, a number of states have groundwater quality standards or use classification systems either relating to drinking water quality or for managing industrial activity impacts on groundwater quality.

Groundwater is a significant source of inflow for streams

Instream flow laws

The purpose of instream flow laws is to set aside water in streams for the protection of wildlife and public health. This body of law responds in different ways to stream water quality and the needs of environmental habitat. Also, instream flows can be required as carriage water to ensure that downstream water right owners get their entitlements of water. Taken together, this yields three reasons to regulate instream water: to provide dilution water for wastewater, to provide enough water for habitat, and to deliver water supplies to downstream users.

Instream flow law is mainly at the state level. States may require that minimum flows remain in the stream to dilute wastewater returns and that minimum levels be left in the stream for fish and wildlife.

While there is no federal instream flow law, other federal laws can be used to protect instream flows. For example, a permit holder might be required to bypass flows around a federally permitted lake to ensure that fish in the stream have enough water.

Interbasin transfer law

Interbasin transfers (IBT), or export of water from *basins-of-origin*, are contentious because no basin wants to lose water to another. In an interbasin transfer, water is removed from its natural watershed and transferred to another basin, usually on a permanent basis. This can have the effect of permanently changing the economy and ecology of both the basin-of-origin and the receiving basin.

Instream flow law works to provide environmental flows

Under the appropriation doctrine, IBT is generally legal. A landmark case in Colorado (*Coffin v. Left Hand Ditch Company*) set a precedent for other situations. However, IBT remains contentious in spite of that ruling. IBTs are also contentious in riparian states, where they are usually handled under the rules of a permit system or the authority of the state government. Proposals for them bring to light uncertainties about the legality of IBT, possible injury to landowners, ownership and conveyance rights, possible environmental impacts, and remedies.

Public utility commission law

State law empowers public utility commissions (PUCs) to regulate the costs of water service for some utilities. These commissions, where they are concerned with water at all, regulate only private water companies. The public is largely ignorant of whether they are receiving the most cost-effective water supply service possible. Electric, gas, and telecommunications utilities have their rate decisions made public, and so comparisons of costs are easier for the public to make. The National Regulatory Research Institute (1983) publishes reports on how states regulate water utilities.

Water utilities and the water industry have lower profiles with PUCs than do the electric, gas, and telecommunications industries. This effect is also apparent in the actions of PUCs, which are dominated by the regulatory scenarios of other public services.

Local laws

Local law mainly occurs as ordinances, such as for stormwater control, floodplain land use, water use restrictions, industrial pretreatment rules, and, in the case of local districts, groundwater use restrictions.

Water use restrictions

Local water use restrictions may be imposed during drought or other emergencies. Authority for these varies and depends on systems of water allocation law. For example, under the appropriation doctrine, local governments cannot impose any restrictions on water users who own water rights. However, they can impose restrictions on customers who draw water from a central system. Growing populations in urban regions may trigger water use restrictions as a result of overtaxed local supplies.

Stormwater ordinances

Stormwater programs, which are typically implemented by local governments, can have significant TWM effects to control hydromodifications. Their authorities are not always clear because stormwater involves goals for flooding, water quality, and land use. Local government programs may include stormwater standards, subdivision regulations, stormwater quality, erosion control, and land quality. These promote stream restoration, greenbelt construction, recreation, and environmental education.

The legal basis for governmental regulation of stormwater is in state constitutions and local charters that authorize cities to improve the health and welfare of citizens. A few states, such as Pennsylvania, have stormwater statutes. Maryland also has an active program, and Florida's water management laws extend to stormwater.

Locally set stormwater standards might set return periods and levels of service for stormwater systems. Subdivision regulations might impose standards and requirements on developers. Related to them might be development standards that impose requirements for amenities such as greenbelts, walkways, and ponds. The stormwater quality program will respond to CWA requirements.

Three basic legal doctrines occur in the law of drainage and hydrologic modification: the common enemy rule, the natural flow rule, and the reasonable use rule (Goldfarb, 1988). Under the common enemy rule, you can do anything to protect your property, regardless of how you affect your neighbor. The natural flow rule is the reverse: you must not change anything that would affect natural flows. The reasonable use rule is more practical. Under it, you may modify your land, even if you affect your neighbor, but there is a test of reasonableness. This rule recognizes that development will occur but that there is a community obligation to work together to accommodate it. Regulations such as required detention storage are examples of reasonable use doctrine approaches.

The Clean Water Act provided authority to regulate stormwater quality. USEPA issued rules in 1984 that stormwater dischargers had to apply for permits. This attempt to require permits failed, but the Water Quality Act of 1987 Amendments mandated new controls and deadlines. In 1990, USEPA issued further stormwater rules, and local governments are facing deadlines that they have not experienced before. Although the 1990 regulations seem old by now, many uncertainties remain as to how they will be enforced.

In many ways, stormwater law is strongly linked to environmental goals

Regulation in the water industry

The process of regulation is how environmental laws are implemented and administered. The term means to control behavior in accordance with a rule or law, and is aimed at protecting the public interest where private markets do not. Regulators are an important part of the water industry, and they enforce rules about health and safety, water quality, fish and wildlife, quantity allocation, finance, and service quality.

Regulations implement laws, and the Administrative Procedures Act (APA) was passed in 1946 to give coherence to rule-making (Gifis, 1984). For example, the Safe Drinking Water Act is a law passed by the US Congress and signed by the president. The Lead and Copper Rule is a regulation issued by USEPA under the authority of the SDWA.

Water industry regulation began with water allocation systems in the West, which was then followed by public health laws related to drinking water. Now it has been extended to environmental issues such as endangered species. Finance is not regulated much, other than for private water companies under state public service commissions. Service quality is regulated indirectly through other programs. For example, water pressures must be kept high enough for fire protection, which is regulated under design codes that respond to insurance requirements.

Regulatory programs should follow the principle of not having the fox guard the chicken coop. This recognizes that persons should not be expected to regulate themselves. On the other hand, the same agencies that write the rules enforce them, so regulators need oversight as well. These are examples of why the separation of powers is required in government.

The regulatory arena is where conflicts over business versus environment are worked out. In this sense, regulation is a *coordinating mechanism* for the water industry. Interest groups push their agendas through regulations and laws. Each sector of the water industry has its own regulatory programs, based on the Safe Drinking Water Act, the Clean Water Act, stormwater rules, floodplain regulation, hydropower licensing, and environmental regulations.

A regulatory program must have an enforcement mechanism to be taken seriously. Most of the experience in the water field is from enforcement of the Clean Water Act, which gives authority to USEPA to enter and inspect premises, review records, test monitoring equipment, and take samples. USEPA can issue compliance orders or take action in civil courts.

Like officials at a sports event, enforcement staff should know the rules well. Officials must have reliable information to base decisions on. Enforcement officials should try to obtain compliance before levying penalties. The system of enforcement must be efficient. Enforcement must be fair, and appeal panels must be available to provide due process as well as to back up the regulatory goals.

In regulatory control, do not have the fox guarding the chicken coop

The total picture of water industry regulation is a mixture of federal, state, and local laws and regulations that govern water service providers and individual water users. Because much of water service is by local government, the regulation comes from federal laws implemented by state agencies. Other regulation is informal, through the political process. For example, rate-setting by local governments normally requires no approvals, whereas rate-setting by private utilities is regulated by public service commissions.

Regulatory programs follow a set sequence in their development and implementation:
- Identification of problems
- Formulation of laws and rules
- Development of rules and programs to administer them
- Staffing, budgeting, and implementation of programs
- Monitoring and enforcement programs
- Systems for appeal of penalties and rulings
- Arrangements to review and modify laws and rules

Calls for regulatory *relief* and regulatory *reform* are common because people and businesses don't like being regulated. However, regulation is a price to pay for civilized society. The challenge is to regulate enough but not too much. Regulation seeks to apply law to control behavior in the public interest, but defining the public interest is an elusive goal.

Roles of courts

It is surprising to note how much the third branch of government—the justice system—is involved in water resources management. The justice system involves federal, state, and local courts, as well as the administrative law system. While the main part of water law is statutory, much of it is *case law*, where complex situations have been tried and precedents have been set. Attorneys search hard for cases to prove their points and to build arguments based on precedents.

Lawsuits are used to gain decisions about complex issues. When an action gets to court, it means the voluntary, coordinated approach has broken down, and court decrees and decisions may take the place of agreements and programs.

> A surprising number of water decisions are made by courts

International water laws

For the most part, water laws are distinct within countries, but international water laws apply to situations such as flows across national boundaries and use of shared waters, such as the oceans. The United States is fortunate in that disputes between states can be settled by the Supreme Court, but when two sovereign nations have a water dispute, they usually lack an arbitrating authority. In some cases, conflicts and even war can result from water conflicts. United Nations agency programs offer avenues for nations to at least discuss their water issues, and forums such as international associations and world courts are available to help resolve disputes, should the nations choose to use them.

The European Union is developing a common approach to water quality law within the Water Framework Directive. It seeks to normalize water quality laws across the nations, much as the US Clean Water Act did for water quality law among the states.

Summary points

- Law and regulations are needed to implement TWM. In addition to not complying with regulations, problems are caused by everyone acting independently and not working together and by myriad small-scale and individual actions that escape regulatory oversight and cause unsustainability.
- Examples of legal issues: There can be too many boats on a lake, too much fertilizer in streams and ponds, too much sediment from construction activities, too many unregulated industrial discharges, lack of adequate wastewater treatment, and many other such problems.

- The upside of laws and regulations is that they provide tools to control harmful or inequitable activities. The downside is that they may place management systems on a rigid autopilot system and block the dynamic process of TWM. Also, laws and regulations work better with large and well-funded players but not so well for the many small players.
- Regulatory controls and law enforcement are required to make TWM work because stewardship alone cannot solve all water problems. Water law includes a body of legal instruments and rules that comprise a system of regulatory control of each type of water decision (capital, operating, regulatory, and financial).
- Excessive reliance on regulations can prevent the TWM concept from seeking a dynamic process that adapts to changing conditions and local and regional variations.
- Laws and regulations can go beyond pure control and provide helpful management instruments to aid planning and coordination.
- Water law refers to all levels and types of law that affect the way water is managed. Regulations are rules or directives that are issued by administrative agencies and backed by the authority of law. The water industry is regulated for: health and safety; water quality; fish and wildlife protection; quantity allocation; finance; and service quality. In some cases the regulation is formal, through government agencies, and in other cases it is informal, such as through self-regulation via the political process.
- The Safe Drinking Water Act is the main federal act that regulates water supply utilities and applies health-based standards to protect drinking water against a range of threats. It provides a framework for USEPA, states, tribes, water systems, and the public to work together to provide safe water. Along with the Clean Water Act, it provides the basis for the multiple barrier protection of public health.
- State laws regulate surface and groundwater allocation, instream flows, interbasin transfers, and public utility commission actions. Water allocation law is especially important in source water supply.
- Local laws and regulations control stormwater, floodplain use, water use restrictions, industrial pretreatment, and, in many cases, groundwater use. With drought and water scarcity, recent years have seen more emphasis on water use restrictions and conservation.
- Federal, state, and local courts, as well as the administrative law system, are sometimes used to resolve disputes in water resources management. When an action gets to court, it means the voluntary, coordinated approach has broken down, and court decrees and decisions may take the place of agreements and programs.

Review questions

1. Explain how water management would differ under management systems based on voluntary stewardship, economic incentives, and regulatory command and control.

2. Give examples of how laws and regulations control decisions in capital, operating, regulatory, and financial business processes.

3. Explain how laws and regulations go beyond command and control to provide management instruments that aid in planning and coordination of water actions.

4. List types of regulation in the water industry and give examples.

5. Explain the main categories of federal, state, and local law that affect TWM actions.

6. How is the court system involved in water management actions?

References

Barrett, Paul M. 1994. Justices Allow Wide State Regulation of Hydroelectric Projects of Utilities. *Wall Street Journal*, June 1.

Corbridge, James N. Jr., and Teresa A. Rice. 1999. *Vranesh's Colorado Water Law*. Rev. ed. Niwot, Colo.: University Press of Colorado.

Federal Emergency Management Agency (FEMA). 2003. *The National Flood Insurance Act of 1968 and Flood Disaster Protection Act of 1973*. http://www.fema.gov/pdf/nfip/floodact.pdf Accessed December 1, 2003.

FEMA. 2004. Chronology of Flood Laws. Unpublished document. Washington, D.C.

Getches, David H. 1990. *Water Law in a Nutshell*. St. Paul, Minn.: West Publishing Co.

Gifis, Steven H., 1984. *Law Dictionary*. 2nd ed. New York: Barron's Educational Series, Inc.

Goldfarb, William. 1988. *Water Law*. 2nd ed. Chelsea, Mich.: Lewis Publishers.

Graham, Andrew G., Wade E. Hathhorn, Robert L. Wubbena, Les K. Lampe, and Neil Grigg. 1999. *Managing Constraints to Water Source Development*. Denver, Colo.: AwwaRF.

Grigg, N., L. MacDonnell, J. Salas, D. Fontane, L. Roesner, C. Howe, and M. Livingston. 2004. *Integrated Water Management and Law in Colombia: Institutional Aspects of Water Management, Allocation of Uses, Control of Contamination, and Urban Drainage*. Fort Collins, Colo.: Colorado State University.

National Regulatory Research Institute. 1983. *Commission Regulation of Small Water Utilities: Some Issues and Solutions*. Columbus, Ohio: NRRI.

Rice, Leonard, and Michael D. White. 1987. *Engineering Aspects of Water Law*. New York: John Wiley & Sons.

University of New Mexico School of Law. 2003. *Federal Wildlife and Related Laws Handbook*. Center for Wildlife Law, Institute of Public Law. http://ipl.unm.edu/cwl/fedbook/fedpower.html. Accessed December 5, 2003.

US Environmental Protection Agency. 1999. *Understanding the Safe Drinking Water Act*. EPA 810-F-99-008.

USEPA. 2003a. *Laws and Regulations.* http://www.epa.gov/region5/water/cwa.htm. Accessed October 17, 2003.

USEPA. 2003b. *Introduction to the Clean Water Act.* Watershed Academy Web. http://www.epa.gov/watertrain/cwa/. Accessed December 5, 2003.

USEPA. 2003c. *Why Set Secondary Standards?* http://www.epa.gov. Accessed December 6, 2003.

US General Accounting Office. 1994. Water Pollution: EPA Needs to Set Priorities for Water Quality Criteria Issues. GAO/RCED-94-117.

CHAPTER 10

POLITICAL AND INSTITUTIONAL OBSTACLES TO TWM

When we hit a sticking point in the water business, we may say, "It's just politics!" Academics might say it this way: "That's an institutional issue." Actually, there is a lot more to it than politics, and in this chapter I explain the issues that block TWM's success. The term I use to lump them together is *institutional issues* but, if you prefer, you can call it politics, culture, incentives, or some other name.

Unless TWM, or some other method of addressing the big water-management picture succeeds, we cannot achieve sustainable development. However, a number of glitches block its way and stand in the way of sustainability. The good news is that positive incentives can overcome some of these problems, but to create the incentives, we need a good understanding of why water management actions fail.

When things do not go well in the water arena, people may find it hard to explain. Instead of "politics," we might blame the "government." There are many ways to express the same frustration, like: "It's the bureaucracy," "It's all the red tape," or "It's the system." A lot of these are in the form of excuses, like, "It isn't know-how that gets you ahead here, it's the know-who." There is also a note of resignation, such as saying, "That's par for the course" or "That's just the way it is."

Sound bites like these communicate many things to us, but they do not explain the issues very well. Let's take a closer look at them. Figure 10-1 gives a graphic display of factors that you hear people complain about, such

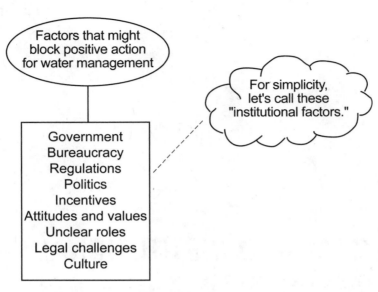

Figure 10-1. A few "institutional factors"

as "government," "regulations," or "attitudes." These are all important, but they give us only a partial picture of the factors that explain how society adjusts to pressures and strong forces.

This chapter explains the institutional factors that affect TWM. In addition to politics and government, it addresses organizations, laws, regulations, authorities, incentives, roles and relationships, and cultural factors (Grigg, 2005). After explaining how these work, I present a *gap analysis* to identify how institutional factors block TWM solutions to problems and I suggest how the functioning of TWM can be improved in the real world.

An explanation of institutional factors

Institutional factors determine how things really work, regardless of what the rulebook says. They are the societal forces and structures that determine patterns of behavior. Formal programs go on the books, but institutional factors determine the outcomes in society.

Sometimes our explanations of institutional factors may seem abstract and theoretical, but it's well to define them in general. The word *institution* is derived from "to institute," which means to establish something. You can say, for example, let's create a new school district, which will become an "institution." A law becomes an institution. We could institute a new habit to make drinking tea instead of coffee an institution. If you established a baseball league, it would become an institution. These illustrate habits and organizations as institutions. If a family's teenage son is constrained by rules and incentives established by his parents, then the rules and incentives are family institutions. Examples of institutional factors that apply to water management are given in Table 10-1.

Table 10-1. Institutional factors relating to water management

Institutional factor	How applied to water management
Organizations	Utilities, regulators, and other players, large and small. USEPA and the California Department of Water Resources are both institutions.
Laws, regulations, and authorities	Water laws.* These are usually laws, but authorities can be mediators, politicians who decide a policy, etc. We could say the Office of the River Master of the Delaware River is an institution.
Incentives	This main category applies across the board. An example could be that people who live in Tucson have an incentive to save water because of the institution of a high water rate.
Roles and relationships	Roles mandated or needed and relationships among players in the water game. We could say that the ditch rider has important roles in harmonizing water diversions and must have good relationships with the farmers.
Culture	How business is really done. We could say that in the Lower Platte Valley, farmers gather at the XYZ café to talk over water management, and these meetings are an institution.

* see chapter 9

Examples of institutional problems

I have found that institutional factors are difficult to explain, and examples may help. Recently, in my class on water resources management at Colorado State University, we were discussing how a river basin should have an integrated plan, a coordinating commission, a set of rules, and other arrangements. A foreign student was returning to her country for research on regional water planning, and I suggested that she interview stakeholders to find out their roles and responsibilities. She replied, "Oh, these are all described in the law." We then discussed how written law is not always reality and that we may not even know why this occurs. A local saying that often brings nods and laughs at water user meetings here explains the point further: "There are three ways to explain how Colorado water rights are administered. There is the law. Then, there is the way people think it works. Finally, there is the way it really works."

As another example, say a legislator gets a law passed but the institutional arrangements are not in place for its success. To make the law effective, the players in industry and government would have to accept it in reality, and the money would have to be there to enforce it. There might be fatal flaws that keep it from working. In Colorado, the state legislature passed a bill one year to fund a State Engineer's regulatory program from permit fees. The next year, it was repealed. Why? It didn't work because the institutional design did not work.

> In Colorado there are three ways to explain how water rights work. There is the law. There is the way people think it works. Then there is the way it really works.

In another example relating to water supply, an Awwa Research Foundation (AwwaRF) study of water source development found constraints that included laws and regulations; government programs and policies; and institutional barriers in the form of multiple stakeholders, overlapping regulatory jurisdictions, lack of cooperation, and unclear approval mechanisms (Graham et al., 1999).

Many such examples come up in discussions about water resources management. A few more that come to mind are:
- Getting people to value water correctly
- Igniting citizen stewardship of water
- Controlling cumulative effects of land development
- Mitigating nonpoint sources through behavioral change
- Enforcing laws and regulations effectively
- Organizing effective governance and coordination mechanisms
- Reaching agreement on the state of the environment
- Sustaining a motivated and effective water workforce
- Ensuring equity and justice in the allocation of water resources
- Motivating water utilities in their corporate social responsibility
- Fathoming scientific complexity in large-scale water problems

Water institutions

The main organizations and agencies of the water industry are the water service providers (including some government agencies) and regulators. Other organizations with significant influence are water industry associations, large firms, research organizations, and coordinating agencies. Some of these are in the industry's support sector and some are governmental. Taken together, these comprise a vast array of stakeholders that align themselves on the *iron triangle* of the water industry.[1]

---

1 See chapter 12.

Laws, regulations, and authorities

The main laws and regulations of the water industry are explained in chapter 9. They are either enabling or regulatory in character, and they create the legal basis for water management actions. The Clean Water Act and all of its regulations comprise an institutional arrangement. Any kind of rule can become an institution.

The formal authorities of the water industry stem from its laws and regulations. Each water utility has a charter issued either by the government or as a corporation. The main regulatory authorities are the USEPA, state regulators, public utility commissions (PUCs), and smaller regulators, such as floodplain offices.

Formal authorities are established by law, but informal authorities such as coordination mechanisms may be established by agreements. For example, an intergovernmental agreement among water agencies might establish a planning group.

Incentives

Incentives are powerful institutional factors. Positive incentives will make TWM work well. We want to avoid perverse incentives that lead to negative outcomes and work against the objectives of TWM. An example of these is an incentive that leads people to keep the adversarial process going, such as litigation that is part of dispute resolution.

The water industry has many informal incentives, such as avoiding the bad press of a boil-water order or a bond default. These informal incentives are found in other industries as well, for what corporation wants to be in the news because its product was defective or it went bankrupt?

Laws and regulations are formal incentives that the water industry relies on to motivate its customers and other entities to act a certain way. Examples include penalties from permit violations, "polluter pay" effluent fees, water rate structures to encourage conservation, government-subsidized flood insurance, and encouragement for safe water through Consumer Confidence Reports. Laws and regulations are also in place to compel water utilities to meet minimum requirements. This tends toward a cop-on-the-beat arrangement, where water authorities are assumed to violate rules and the regulators are there to catch them. So where does the stewardship ethic come in? It's like a voluntary ethic, similar to a business that might say: "We are good citizens; we only sell green products." Who believes that? So, in TWM, utilities take a proactive approach that acknowledges the limited effectiveness of purely regulatory approaches and leads toward a positive and stewardship-based solution to difficult water issues.

Table 10-2 shows a way to classify incentives in the water sector.

Table 10-2. Classification of incentives in the water sector

Type of incentive	How it works
Performance-based incentives	Managerial tools that reward success in TWM
Avoidance of civil or criminal penalties	Preventing violations of permit conditions
Risk avoidance for financial losses	Avoiding negative financial results in the management of a utility
Avoidance of bad publicity	Avoiding bad publicity about public health and safety
Good customer service	Incentives that reward excellence in utility customer service
Stewardship or value-based incentives	Incentives to exercise environmental responsibility

Roles and relationships

Roles and relationships in the water industry are similar to those in other industries. They begin with the roles of service provider, regulator, and support organization. They are created by formal authorities and by informal relationships, such as those formed by working together in civic associations.

Culture of the water industry

A few realities of the water industry serve to illustrate the culture. One is that people take water for granted and think it should be free or at least low in cost. There are built-in conflicts between water developers and environmentalists because of their incentives, so the culture between them is adversarial. People like water recreation, and our culture promotes it. *Culture* is the integrated result of habits and relationships formed over time. For example, in the water rights example given earlier, informal water sharing is a cultural concept not recognized in the law. Another cultural institution could be a regular arrangement, such as farmers meeting to talk at the same time every day.

Examples of institutional obstacles to TWM

In Table 10-3, difficult problems that confront TWM are listed. Longer explanations follow in discussions of water supply and nonpoint source control as special cases.

Table 10-3. Discussion of problems confronting TWM

Problem facing TWM	Causes
Relentless development and hydrologic modification	As people continually develop and "improve" land, stream networks undergo extensive modification. Development follows the profit motive. In parts of the world, the culture is oriented toward stewardship and in others it is not. How can incentives be developed to sustain stream networks in ecologically sustainable condition?
Organizational limits that limit corporate social responsibility in utilities	Utilities and water agencies respond to incentives and have their own cultures and significant corporate social responsibilities. How do they juggle these with their obligations to offer services to customers at low rates?
Political-to-watershed mismatches and regional disputes	Political boundaries usually do not match watershed boundaries. Competition between regions extends to disputes over water matters. How do we achieve effective watershed management without watershed authorities?
Jurisdictional gaps	Laws and regulations are fragmented, and there are gaps between authorities over sectors and parts of systems. For example, one agency regulates water in a stream and another regulates drinking water systems. One set of regulators deals with water distribution systems and another set with premise plumbing.
Tragedy of Commons	People take care of their property but have no direct incentives to care for public property. What is everyone's interest is in no one's private interest. If a homeowner disposes of a small amount of toxic waste, he may have to drive 10 miles to a hazardous waste disposal facility. He dumps it in a stormwater inlet, reasoning that this small amount hurts no one. He may not know or care about the ecosystem effects.
Equity problems	Equity problems cut across issues. For example, wastewater organizations comply with laws, have relationships with regulators and neighbors, have incentives not to violate standards, and have cultures of care and concern or lack of them. However, an upstream wastewater discharger lacks incentives to protect downstream water users. People seeking interbasin transfers do not consider the needs of the basin of origin. Under the appropriation doctrine of water law, the use-it-or-lose-it feature encourages water waste and discourages water sharing in times of need.
Cost-of-service rates	Rates set at the marginal cost of service do not consider the full societal costs of water and may encourage misallocation.
Government subsidies	Government subsidies distort incentives, as for example in subsidizing irrigation systems.
Scientific complexity	In a complex water quality problem facing an agency, it will be difficult to prove or disprove certain links between management actions and permit violations. Given the difficulty in reaching agreement, how can science produce consensus answers on complex and large-scale scientific problems such as the Everglades or the California Bay Delta?

Continued on next page.

Table 10-3. Discussion of problems confronting TWM, *continued*	
River basin commissions (RBCs) are often ineffective	If RBCs are imposed top-down, they may lack the champions needed to forge cooperation and coordination. The Water Resources Planning Act enabled RBCs, but they have since been dismantled. Some interstate compact commissions have not worked well for the same reasons. Failure of these units can be explained by incentive sets.

Water supply industry constraints

The reality in the water supply industry is that, while the industry debates sustainable development, it must also deliver reliable and safe water services to its customers. Water utilities are concerned with developing and maintaining sources of water and with complying with health, environmental, and safety regulations.

When the concept of TWM was developed by AWWA, a workshop group studied key issues within it and determined that the main issue faced by water supply utilities was managing constraints to new source development. This resulted in a project by the AwwaRF entitled *Managing Constraints to Water Source Development* (Graham et al., 1999).

> **Rising water demands and environmental interest from the public make water source development a challenge**

The central issue of the study was that rising water demands come from the same public that places environmental and other constraints on water source development. TWM is a good vehicle to balance these often conflicting demands. The study considered 10 cases from around the United States, with the results that some places, especially in the West, are experiencing water shortages whereas others, such as Detroit, have enough water but might experience conflicts within their region, such as demands on the Great Lakes.

The study considered legal, institutional, and policy constraints that included laws and regulations, government programs and policies, and institutional barriers in the form of multiple stakeholders, overlapping regulatory jurisdictions, lack of cooperation, and unclear approval mechanisms.

The study group proposed a framework for analyzing constraints, basically a display of two variables compared to two others, of stakeholder preferences and authorizer responsibilities and interests. Eighteen constraint categories were identified. Virtually all of the utilities examined faced some constraints to a degree that affected planning for new sources.

These were the recommendations:[2]
1. Identify key public values and assess their relationship to water supply plan elements
2. Increase emphasis on stewardship of water resources
3. Use an adaptive management framework to manage uncertainty
4. Find common ground between urban and rural communities
5. Consider project elements that achieve benefits for other stakeholders
6. Coordinate planning among multiple utilities within water supply regions
7. Incorporate a 50-year planning horizon in water-resource planning efforts
8. Build internal capacity in terms of negotiation, organizational analysis, and political strategy
9. Emphasize indirect benefits of stewardship activities
10. Emphasize benefits of regional problem-solving among utilities
11. Improve alignment of values among water-resource interests
12. Work at state, provincial, and national levels to improve standing of water supply among authorizers
13. Maintain and increase training opportunities to assist utilities to build capacity in negotiation, organizational analysis, and political strategy

These recommendations focus on political strategy rather than any technical steps. The use of the word *political* is in the sense of making government work well, rather than any electoral politics. These topics are discussed in chapter 4.

Nonpoint source pollution

Institutional obstacles loom large in finding solutions to nonpoint source problems. The issue has been studied extensively since the enactment of the Clean Water Act in 1972 and was evaluated in the Water Quality 2000 study. Water Quality 2000 (WQ 2000) coined the slogan "Society living in harmony with healthy natural systems," which is a way to express the concept of sustainable development. It involved more than 80 public, private, and nonprofit organizations to recommend a national policy for water quality management (Water Quality 2000, 1992; Water Quality 2000 Steering Committee, 1991).

---

[2] The authors listed recommendations 1–8 for utilities and 9–13 for national water supply associations.

This effort was a good illustration of TWM in action. It sought broad representation; a long-range, visionary, and holistic perspective; maximum consensus on national principles; and a focus on water quality with a balanced view and a specific agenda for action. The goals were consistent with an integrated approach to water management. The study found that information provided by USEPA reports is useful but not adequate to assess the condition of the water bodies. Problems revolved around the complexity of aquatic ecosystems, the expense of comprehensive monitoring programs, and the patchwork nature of reporting systems. It found that ambient monitoring was far too limited to assess water quality and that data were incomplete, covering only a fraction of all waters and pollutants. This leads to conflicting reports on the condition of water and ecosystems.

WQ 2000 explained that sources of pollution are driven by decisions about how society lives, farms, produces and consumes, transports people and goods, plans for the future, and acted in the past. How we acted in the past explains problems such as acid mine drainage, polluted groundwater, and contaminated sediments. It did acknowledge progress in the return of game fish to waters that were impaired, but there were still failures in the form of destruction of habitat, fish and shellfish being contaminated, and violations of water quality standards.

For institutional gaps, Water Quality 2000 listed narrowly focused water policy; conflicts between institutions; legislative and regulatory overlaps; conflicts and gaps; insufficient funding and incentives; inadequate attention to the need for trained personnel; limitations on research and development; and inadequate public commitment to water quality.

Water Quality 2000: We pollute by how we live

Narrowly focused policy leads to the easy targets of point source controls and conventional pollutants, rather than overall water quality improvement. Issues that have not been addressed as a result include watershed-based planning, cross-media effects, relationships between water quality and quantity, pollution prevention, and focusing on environmental results rather than statistics like the number of treatment plants.

Institutional conflicts arise because of the federal system of government with its many interest groups. They include questions about the allocation of authority among the key players and conflicts among groups at any particular level.

Legislative and regulatory gaps might be due to uncoordinated regulatory programs. For example, there might be different state and federal standards dealing with the same issue. There might be a conflict in a government program that provides tax deductions for second homes but then allows those homes to be located in environmentally sensitive areas.

Insufficient funding and incentives for water quality programs result from the rapid 1970s ramp-up of the Clean Water Act. Subsidies that built sewer systems and treatment plants have gone away. If the lifetime of those systems is on the order of 20 to 30 years, they now need massive investments for renewal.

WQ 2000 flagged as critical issues reducing the cost of clean water and healthy ecosystems, paying the remaining cost, and allocating funds among competing investments. It also noted that more stringent regulatory standards and sophisticated control equipment require more education and training and that more public commitment is needed for water quality.

Water Quality 2000 identified 12 emerging issues: preventing pollution, controlling runoff, toxic constituents, protecting aquatic ecosystems, multimedia pollution, groundwater, scientific understanding of water quality issues, promoting wise use of resources, setting priorities, providing safe drinking water, managing growth and development, and financing water resource improvements. These are still valid issues some 15 years after the study.

WQ 2000 was a commendable effort and confirmed what other policy makers had been saying: water resources policy is pulled in too many directions, with competing and conflicting players, programs, and priorities. Congress thought that WQ 2000 was instrumental in educating and motivating them, the executive branch, and the public to pursue a more sustainable approach to water that features pollution prevention, individual and collective responsibility, and an integrated watershed approach. Congress is expected to continue its focus on moderate, bipartisan legislation and on searching for the middle ground or consensus on water issues that WQ 2000 recognized as difficult. Examples are wet weather flows, total maximum daily loads (TMDLs; USEPA received 30,000 comments on this policy), and wetlands protection (Boehlert, 2000).

A method for institutional analysis

Given the many possible institutional factors, we need a way to analyze them systematically. Ziegler (1994) presented a generic method based on his definition that *institutional analysis* studies the patterns of human activity in groups, the rules of the game, and how to modify behavior by altering the patterns that direct it. His key questions for an analysis were:

- What goes on in this situation?
- What processes need adjustment?
- What problem-solving know-how is available?
- What ought to go on here?
- What are the impacts of change on other patterns of activity in this institution?
- What are the impacts of change on other institutions?

If you add in the categories of institutions mentioned earlier (laws and controls; authorities and stakeholders; incentives; roles, responsibilities, and relationships; and culture) you can derive a three-step process for institutional analysis (Grigg, 2005):

- Create a conceptual model of how the management and control systems work (What goes on here?). This could be, for example, a flow chart.
- Identify the key issues in each category of institutional element (What processes need adjustment?). This could start with business processes and the obstacles they face in the form of a gap analysis.
- Identify institutional practices that should lead to improvement (What ought to go on here?). This can be a set of recommendations about how to improve a situation.

These questions relate well to the processes of *strategic planning* and *gap analysis*. The process also fits into *systems thinking*, a popular method that includes looking at the big picture or mental model of a situation (Senge, 1990).

Political model of water planning

The combined effects of institutional factors lead to a political model of water resources planning, as shown in Figure 10-2. Back in chapter 4, a rational model was shown at work inside of a political process. This figure shows the effects of political issues and the incentives of players as they unfold over time.

In the political model, we recognize that the resolution of water conflicts usually lies in the legal, financial, and political arenas, rather than in the technical arena. A problem is initially identified through some process of management or politics. The planner will want to know who the stakeholders and decision makers are. The stakeholders need to be involved up front in a substantive way. The process unfolds over time, maybe years.

The features of the process include the problem to be solved, stakeholders, coalitions, goals, strategies, study processes, decision points,

CHAPTER 10   POLITICAL AND INSTITUTIONAL OBSTACLES   247

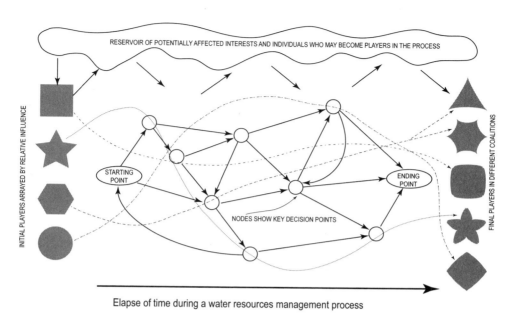

Figure 10-2.   Water resources decision process

and possible outcomes. The stakeholders are arrayed in levels of power or influence in the decision process. The positions of the various actors also may vary over time. The stakeholders array themselves in coalitions or interest groups, and these rise and fall in influence and interest during the process.

Stakeholders enter and leave the overall process during the long time period of many water resources problems and projects. We may not like the fact that not all stakeholders are equal, but it is a reality.

At the far right of the diagram lies a set of possible outcomes. The possible outcomes have variable characteristics, including technical alternatives, institutional alternatives, alternative goal achievement, alternative management arrangements, alternative timing, and alternative location dimensions. Sometimes the alternative outcomes can be related to other outcomes as in inter-sector planning problems.

Along the route to the decision lie numerous crucial decision subpoints that involve some or all of the stakeholders. These can be meetings, reviews, the completion of studies, new developments and surprises, and changed attitudes. In between these nodal decision-process subpoints lie decision subprocesses. Sometimes these seem quiet and inactive, but the committed groups know that crucial matters are under way. For example, influence and power are shifting, and knowledge is building. The decision subpoints may be the steps of the planning process, such as identifying alternatives, but these steps are really complex exercises in themselves.

> **In many cases, the political model of water planning is closer to the real thing than the rational model**

It is important to realize that the processes referred to are carried out in the absence of perfect information. There is a lack of organized information and intelligence about what is going on with allies, neutral parties, and opponents in water resources problem solving.

Some stakeholders will influence outcomes from the beginning. They need clear goals early on so that political strategies can be formulated. They have advantages over others who may decide to "go along" or participate less actively in the decision process. To be effective, coalitions of stakeholders are needed and maximum influence is sought. Environmental organizations have gained the reputation of being focused in their goals in opposing water resources development.

Gap analysis and remedies

Actually, an institutional analysis leads to what some people call a gap analysis, or an analysis of the gap between what should be and what is.

On barriers and gaps in IWRM, Gilbert White (1998) reviewed 50 years of experience and concluded that the barriers were formidable. He focused on difficulty for water organizations to examine their full "range of choice," complexities of large and small projects and actions, and the heavy pressure on environmental systems from the innumerable actions that fly below the radar screen of water management organizations. This was a sage prediction of the kinds of problems outlined in this chapter. The problems can be explained in different ways, and one way to identify gaps, or the difference between what is needed and what exists, as is shown in Table 10-4, which also includes ideas for fixing the problems.

Roles and responsibilities

When you have a systemic issue, such as TWM, it helps to look at it from more than one angle. Table 10-5 shows a list of TWM requisites and identifies the TWM participants who have the main roles. The table shows that utilities and regulators have clear roles, but other roles are distributed among the players.

Table 10-14. Gaps in need and strategies for fixing them

Gap (difference between what is needed and what exists)	Strategy for fixing the problem
Valuing water fully. Getting people to appreciate the value of water and overcoming distortions such as cost-of-service rates and government subsidies. Cost-of-service rate distortions arise from the "it's not my problem" syndrome.	These are rooted in the economic view that externalities can be built into the rate structure. The remedy is good planning at the local level where the externalities are in fact built into the rate structure. Government subsidies require effective political processes to be rooted out.
Improving governance. Improving management of utilities to get more attention for TBL effects. Overcoming obstacles because of mismatches between political and watershed boundaries. Solving regional disputes. Bridging jurisdictional gaps. Organizing effective coordination mechanisms. Organizing joint planning groups.	These problems are tough because of the regional problem and the problem of political boundaries and thus political incentives. Regional disputes might be mitigated by new arrangements to share authorities and tax bases, but these are hard to come by, except when need is clear. Planners lack authority to bring people together. Politicians have authority but may act only when urgent or in self-interest. Politicians who focus on public interest may get worn out with the long durations and hard knocks of water negotiations. Must be counseled to hang in there for long term.
Igniting citizen stewardship. Changing attitudes of citizens to elevate their concerns about water stewardship.	This will have to go beyond "environmental education" to embrace the intangibles of ethic and commitment. Continued work to emphasize our shared stake in the environment is required, along with information to alert citizens about the full range of water issues.
Stopping cumulative effects. Finding ways to halt the slide caused by relentless development and ratcheting cumulative effects. Implementing effective control systems for myriad small effects, such as in nonpoint source discharges.	As there is no formula to show decision makers exactly how everything is connected, getting such a formula and communicating it clearly would be a start.
Improving enforcement. Making enforcement of environmental laws and regulations more effective.	This issue requires the same attention as other law enforcement programs. Enforcement staff should be enabled, the system enforcement must be efficient and fair, and it should work the way it is designed by law. Regulators have their own culture, and ways must be found to adapt their practices with those of utilities in ways to ensure oversight but also cooperation.

Continued on next page.

Table 10-14. Gaps in need and strategies for fixing them, *continued*

Gap (difference between what is needed and what exists)	Strategy for fixing the problem
Improving environmental indicators. Finding ways to measure and report the state of the environment in ways that raise public awareness and create positive action.	This is a challenge because so much information must be packed into the indicators. Having a system of indicators linked to plans and goals will be a starting point.
Building workforce capacity. Implementing the institutional arrangements that build a motivated and effective water workforce.	This is a systemic issue that requires attention to capacity-building from identification of staffing needs through recruitment and retention. (See Grigg, 2006.)
Making water management more equitable and just. Improving equity in allocation of water resources, upstream-downstream effects, and interbasin transfers.	Equity as an issue must often be enforced by the regulatory and judicial systems. A review of their effectiveness would be a starting point for analysis of equity and justice, and the corresponding public support that results from them.

Table 10-15. TWM roles and responsibilities

TWM requisite	Main role and responsibility
Policy and legislation	Government (three branches)
Sustainable water services	Utilities
Regulation	Regulators
Control/compliance with NPS rules	All NPS dischargers
Making hydromodifications sustainable	All modifiers
Stewardship and education support	Environmental education a shared responsibility; knowledge delivery from knowledge sector and press; association support to create solidarity; advocacy and civics from citizenship development programs in society
Technology and intellectual support	Private sector and think tanks of government and industry

Summary points

- In the institutional arena, the challenges are formidable like they are in other complex sectors. Bok (1996) explained how hard it is to succeed in complex arenas such as environmental protection: "Tasks such as . . . protecting the environment . . . call for other skills—complex planning, building public consensus, coordinating many organizations and agencies, cooperating with community groups, creating efficient bureaucracies. Faced with challenges of this kind . . . certain failings have repeatedly cropped up in field after field of American public policy, hindering the country's efforts to achieve important national goals."

- Former president and war leader Dwight Eisenhower expressed how people don't really want to think about complex situations, and they look for clear solutions. In a military context he wrote, "It is a characteristic of military problems that they yield to nothing but harsh reality; things must be reduced to elemental simplicity and answers must be clear, almost obvious" (quoted in Axelrod, 2006). The problem with TWM situations is that they do not lend themselves to simple answers: they are complex, hard to figure out, and take place over a long time. For these reasons, an approach is required that is different than the one that works in the military. The approach will involve many stakeholders and take on the nature of a sustained campaign.
- Institutional factors determine how things really work in the water industry and govern what can be accomplished through TWM. These include politics and government, organizations, laws and regulations, authorities, incentives, roles and relationships, and cultural factors.
- Water service providers and regulators are the main players in the water industry, which also includes industry associations, large firms, research organizations, and coordinating agencies. The formal authorities of the water industry stem from its laws and regulations. The laws and regulations of the water industry create the legal basis for water management actions. Positive incentives are needed to make TWM work well. Roles and relationships in the water industry define how service providers, regulators, and support organizations work together. The culture of the water industry determines how people value water and work together to solve problems.
- Institutional factors require a political model of water resources planning. In it, resolution of water conflicts occurs in the political, legal, and financial arenas, rather than in the technical arena.

Review questions

1. Using institutional factors as a background, explain why it is hard to implement sustainable water resources management.

2. Explain why water problems are different from business and military problems in terms of the complexity of solutions.

3. Explain what is meant by the culture of the water industry and give examples of how it works.

4. When resolution of water conflicts occurs in legal arenas, is this beneficial to society and a good use of the courts, or does it represent some kind of failure in water management? Give examples to support your answer.

References

Axelrod, Alan. 2006. *Eisenhower on Leadership*. San Francisco: Jossey-Bass.

Boehlert, Sherwood. 2000. Speech to Water Environment Federation. Georgetown University Conference Center, Washington, D.C. Boehlert was Chairman of the House Water Resources and Environment Subcommittee.

Bok, Derek. 1996. *The State of the Nation: Government and the Quest for a Better Society*. Cambridge, Mass.: Harvard University Press.

Duncan, A. 1998. *A Snapshot of Salmon in Oregon*. EM 8722. Oregon State University Extension Service. http://eesc.orst.edu/salmon/human/cumulative.html. Accessed November 22, 2006.

Graham, Andrew G., Wade E. Hathhorn, Robert L. Wubbena, Les K. Lampe, and Neil Grigg. 1999. *Managing Constraints to Water Source Development*. Denver, Colo.: AwwaRF.

Grigg, N. 2005. Institutional Analysis of Infrastructure Problems: Case of Water Quality in Distribution Systems. American Society of Civil Engineers, *Jour. Management in Engineering* 21 no. 4: 152–158.

Grigg, N. 2006. Workforce Development and Knowledge Management in Water Utilities. *Jour. AWWA* 98 no. 9: 91–99.

Senge, Peter M. 1990. *The Fifth Discipline: The Art and Practice of the Learning Organization*. New York: Doubleday Currency.

Water Quality 2000. 1992. *A National Water Agenda for the 21st Century*. Alexandria, Va.: Water Environmental Federation.

Water Quality 2000 Steering Committee. 1991. *Challenges for the Future*. Interim Report, June, Water Pollution Control Federation, Washington, D.C.

White, G.F. 1998. Reflections on the 50-Year International Search for Integrated Water Management. *Water Policy* 1: 21–27.

Ziegler, John A. 1994. *Experimentalism and Institutional Change: An Approach to the Study and Improvement of Institutions*. Lanham, Md.: University Press of America.

CHAPTER 11

ENVIRONMENTAL STEWARDSHIP, ETHICS, AND EDUCATION

"Before I hear how much you know, I want to know how much you care." This saying captures a main point of this chapter. How much you care depends on your values, and these are even more important than knowledge. Author Stephen Covey (1991) wrote about how each person's life needs a "north star" to direct it toward right goals. This north star includes a lot of values, and values shape our attitudes. Stewardship is a value, an attitude, and an ethic.

> **TWM is about stewardship, which requires ethics, education, and responsibility**

TWM is about stewardship. It is "the exercise of stewardship" and it "requires the participation of all units of government and stakeholders." Both of these statements emphasize our shared responsibility to take care of a limited and precious resource. This chapter is about stewardship, all aspects of it. It covers individual stewardship, organizational responsibilities, environmental ethics, and environmental education.

What change is needed?

As we saw in last chapter's gap analysis, valuing water is a big challenge to TWM. This cultural and attitudinal problem requires change as follows:
- Individuals need to value conservation of resources and practice

sustainability in water use and management because TWM is the water element of sustainability;
- Organizations and their leaders must see the entire TWM picture, learn why it is important, and commit to it in policy and action; and
- Small nonpoint dischargers and hydromodifiers need to understand their impacts and be enabled to practice stewardship in a cost-effective manner.

These changes require new attitudes and incentives to be formed through the learning process so that we can become a learning society in our approach to water management. If citizens and leaders alike appreciate and celebrate the value of water, including small headwater systems, they will be committed to stewardship.

About stewardship

Stewardship means taking care of something, such as our common heritage of water. It is the application of the Golden Rule to water systems. Without stewardship, the cumulative effects of development will degrade water systems and the ecosystems that depend on them. This will threaten life, health, and the environment. Regulations and government programs are not enough; stewardship is required.

TWM depends on stewardship because the incentives for the participants do not go far enough to reach sustainable development. Incentives drive water managers toward least-cost solutions and prevent hydromodifiers and nonpoint dischargers from using sustainable practices. This leads to the cumulative effects that are so damaging to sustainability.

Stewardship is the Golden Rule for water systems

What keeps us from being good stewards? Is it lack of knowledge, lack of incentives, or indifference? Whatever the answer, in the final analysis stewardship depends on understanding, commitment, enablement, and action. Social capital is a big part of stewardship. Society can work together, have a sense of association and shared values, and work toward unified approaches to solving shared problems.

Individual stewardship

Every person has a role in water stewardship. If each player takes on his or her own role and reaches out to do a little more than is required, much of the problem will be solved. The challenge is in getting the players to take on their own responsibilities and to "go the extra mile."

The starting point in individual stewardship lies with knowledge, values, and attitudes. While people today are more aware of environmental issues than they were a few decades back, as shown by support for the "green" movement, they may not think about the impacts of the choices they make. As an example, in an affluent mountain community in Colorado I noticed algae forming on the bottom of nearby small and large streams. Incredibly, people were watering and fertilizing lawn grass just like they do in a big city. They came to the mountains for the natural beauty, and they were inadvertently spoiling it because they didn't understand that adding fertilizer to lawns degrades their beautiful streams.

There are a lot of ways to exercise individual stewardship of water. Practicing conservation, not polluting, and paying attention to environmental needs for water are good ways to do it. Wildlife advocates, farmers, and fishers can be good stewards just as environmentalists can.

TWM is all about individual stewardship.

Corporate stewardship and social responsibility

Everyone should practice individual and citizen stewardship, but some people have more influence on sustainable water systems than others do because of their leadership positions in the water industry. They can practice corporate stewardship by practicing good water management in the systems they control and by undertaking corporate social responsibility (CSR), which includes practices all the way from teaching kids about the environment to cleaning up streams on Saturdays.

CSR is good business and today's youth, the millennial generation, appreciate it. According to a corporate citizenship survey, these young people appear civic-minded and some 69 percent will consider a company's social commitment when deciding where to shop (Cone Inc., 2006).

The idea behind CSR is that businesses should be sensitive to all stakeholder needs, including environmental needs. CSR is linked to sustainable development in the sense that companies consider social and environmental as well as financial consequences.[1] It also extends to worker health and safety, safe and responsible products and services, doing good deeds in communities, and the like.

Today's youth believe in corporate social responsibility

While CSR sounds good, it can be controversial. Some people think it and environmental ethics go too far and do not consider the needs of people enough. Fringe groups can include extremists that may become

---

1 See chapter 5 on the Triple Bottom Line

"eco-terrorists" or activists who want to ruin corporations because these groups allege they fall short in CSR.

Some people think CSR is "anti-capitalism" (Henderson, 2002). Milton Friedman, the Nobel Prize–winning free-market economist, thought it socialistic (Manne, 2006). The rationale behind this thinking is that business's responsibility is to make a profit so that the public's shares will go up in value. This seems to be a version of the "it's not my problem" syndrome.

AWWA supports the notion of CSR. It has partnered with the Nature Conservancy and other organizations to create a Blue Water Award that would recognize success in balancing public health and safety with protecting freshwater ecosystems and meeting water supply needs (Richter, 2007). The award would recognize criteria such as: water quality; source water protection; environmental (instream) flows; efficiency and conservation; and integrated water resources management. This is strongly aligned with the goals of TWM.

Environmental ethics

Environmental ethics, which studies relationships between humans and the environment, addresses our responsibilities. As ethics means a system of moral values and a study of right and wrong behavior, environmental ethics is the field that studies right behavior toward the environment. Environmental ethics is a complex field with topics that range across philosophy, religion, law, and related fields. At Colorado State University, the Department of Philosophy has offered a special course on the subject (Rolston, 1988).

Environmental education

Since the environmental movement began, and in particular after the first Earth Day in 1970, there has been a large effort toward *environmental education,* defined as any organized school or public education effort to teach about how natural environments function and how to manage behavior and ecosystems to live sustainably. This effort has done a lot to raise national awareness of the need to respect and protect the environment.

After the launch of Earth Day and the passage of the National Environmental Policy Act (NEPA), there were incentives for environmental education programs for K–12, universities, government agencies, and corporations. Environmental education also grew at the same time that we saw a great rise in the numbers of nonprofit organizations in the country.

Today, environmental education is practically a profession and has its own associations, such as the North American Association for Environmental Education.

In the field of water education, which is a central part but not all of environmental education, a number of programs have been initiated. One well-known program is Project WET, or Water Education for Teachers. Project WET (2007) was started at Montana State University and focuses on citizen awareness, knowledge, and stewardship of water resources through teaching aids and education programs. Its core beliefs show a balanced approach to water management and illustrate how environmental education can support TWM:

> **After Earth Day in 1970, there was a big increase in the number of environmental education programs**

- Water moves through living and nonliving systems and binds them together in a complex web of life.
- Water of sufficient quality and quantity is important for all water users (energy producers, farmers and ranchers, fish and wildlife, manufacturers, recreationists, rural and urban dwellers).
- Sustainable water management is crucial for providing tomorrow's children with social and economic stability in a healthy environment.
- Awareness of, and respect for, water resources can encourage a personal, lifelong commitment of responsibility and positive community participation.

The need for environmental education is implicit in a number of federal programs, and was recognized formally in the National Environmental Education Act of 1990 (US Congress, 1990). In the act's preamble, Congress found that "threats to human health and environmental quality are increasingly complex, involving a wide range of conventional and toxic contaminants in the air and water and on the land" and that "there is growing evidence of international environmental problems, such as global warming, ocean pollution, and declines in species diversity, and these problems pose serious threats to human health and the environment on a global scale."

So environmental education includes topics across the spectrum from understanding biology and wildlife to global issues such as greenhouse gases. In that sense, water management involves a subset of the universe of environmental issues, and understanding the need for water stewardship requires more than basic awareness; it requires understanding of the details and acceptance of each person's role in sustaining our shared water environment.

Environmental education is not always appreciated. Some people see it as the "saved" trying to "convert" the unsaved. Another problem is that it can be value-laden, so who decides the content and tone is at issue. Just as there is a range of ways to be an environmentalist, so too there is a range of ways to do environmental education.

Roles and responsibilities

An exciting aspect of environmental education is that everyone has a role. For this reason, it offers us a way to get involved in TWM through the back door, so to speak. By this I mean that while not everyone can turn the spigot on flow from a big dam, we can all participate in learning and discussions about stewardship.

In some ways, education is a great equalizer. In the United States, we have a core belief that anyone can rise to the top, and being at the top can be defined in different ways. Education is an enabling mechanism to achieve social goals, and environmental education is an enabling mechanism to achieve environmental understanding and stewardship.

> "There is growing evidence of international environmental problems . . . [that] pose serious threats to human health and the environment on a global scale."
> —US Congress, 1990

Just as our democracy balances values through our economy and social institutions, we balance the dissemination of knowledge and values through a dispersed education system. We have learned that government alone cannot solve all of our problems, but neither can business, so there needs to be a *third way*. This third way involves cooperation among the sectors, with each finding its role through cooperation among government, business, the legal system, and citizen-based initiatives. In the 1830s, the French writer Alexis de Tocqueville observed that our young democracy exhibited a "can-do" citizen attitude, not like old Europe with its tight control by monarchies, clergy, and the nobility. That same principle applies to environmental education today as a mechanism to infuse values and knowledge.

Government has an important role in environmental education, and its laws and programs provide policy guidance as well as some funding to initiate programs. Schools also have critical roles, from K–12 through university education. As a matter of fact, a number of academic subjects can be classified as environmental education—biology, geography, and some physics, for example. The media plays a part in environmental education as well, through publications such as *National Geographic, Outside*, and other magazines, newsletters, and Internet blogs on topics ranging from fly fishing to landscape painting.

As de Tocqueville observed, Americans figured out ways to solve all kinds of problems through associations, so we have a strong tradition of private volunteer activity. After the 1960s, there was a rapid increase in the number of private volunteer organizations (PVOs) to work in many fields such as charity, philanthropy, environmental activism, welfare, and children's needs. Closely related to PVOs are nongovernmental organizations, or NGOs. While the terms are related, there can be significant differences between NGOs and PVOs. An NGO is more program-oriented, like the Red Cross, whereas a PVO works more through volunteers, like a local charity.

> **Environmental education equips us with understanding and a commitment to stewardship**

Many of these organizations work in one way or another in environmental and water education. For example, California's Water Education Foundation offers an impressive array of publications, seminars, tours, and other outlets to inform citizens about state water affairs. Colorado has started a foundation modeled after California's. The American Water Works Association (AWWA) publishes water booklets for teachers and students, and its affiliate, Water for People, reaches out to aid developing countries with water projects. The Rotary Club, a worldwide organization with chapters in many countries, is making water outreach one of its priority programs.

PVOs have advantages over government in the field of environmental education in that they represent stakeholders in education, motivation, encouragement, and volunteer efforts. Government's power lies in different realms, such as to pass laws, regulate, and appropriate tax money. Derek Bok (1996) explained that the nation is finding new ways to solve problems in the Internet age through cooperation between government and NGOs.

> **Former Harvard president Derek Bok notes that the nation is finding new ways to solve problems through cooperation between government and NGOs**

Government can require people to do things, but there are gaps between what is needed and what they are required to do, and gaps between what we are required to do and what we actually do. So what we ought to do and what we actually do should be governed as much by social norms and ethics as they are by law.

Businesses have important roles in environmental education as well, and their programs can represent one line of their corporate social responsibility.

Environmental leadership

Environmental education offers opportunities for leadership in shared programs that can forge consensus about stewardship. If every leader with influence over water resources would embrace sustainability and TWM, it would transform water management. These leaders could say that our shared future requires that we commit to sustainable water management, following the principles of TWM.

In many ways, environmental education is about civics, which is an important pillar of democracy and everyone's responsibility. Water education is linked to civics education in a number of ways. It supports broad civics education by helping people understand society's values and explaining the links between water management, civic life, the economy, and social welfare. It can foster citizen trust in government by helping people feel better about paying for water and environmental protection. It can foster public spirit and a sense of community by helping people to cooperate, by explaining why water management involves the whole community, and by identifying shared solutions to environmental problems.

TWM is all about leadership and civics

When the public is involved in problem-solving, it fosters participation in democracy by explaining the impacts of actions and organizing public involvement in planning and decision-making. It promotes conservation and security by explaining personal responsibility in conservation of resources and environmental protection. It also promotes social and environmental justice by helping people understand the rights of citizens to equity in access to public services and environmental resources.

Requirements for environmental education and ethics

At the end of the day, the question is one of action and how to develop mechanisms to promote and sustain TWM. Environmental education and ethics, along with corporate social responsibility in a broad sense, are valuable supporting concepts to marshal the knowledge and commitment needed to advance TWM.

TWM and its cousins, environmental education and ethics, require knowledge and values. It is interesting to examine the TWM definition to

Table 11-1. Knowledge and value requirements of TWM elements

TWM element	Knowledge requirement	Value requirement
Exercise stewardship of water resources for greatest good of society and the environment	Ethics: defining what stewardship means and what "good" means in the context of TWM	Accepting that stewardship is important and everyone's responsibility and deciding on fair treatment to society and the environment
Supply is renewable, but limited and should be managed on a sustainable use basis	Hydrology and ecology: knowledge of water balance and environmental sustainability	Accepting that the supply is limited and committing to sustainable use
Encourage planning and management on a natural water systems basis	Water management: education to define natural systems basis	Deciding to plan and manage on a natural systems basis
Through a dynamic process that adapts to changing conditions; balance competing uses of water	Political science: development of balancing process that is dynamic and adaptable	Committing to adaptive management and mechanisms to enable balancing
Efficient allocation that addresses social values, cost effectiveness, and environmental benefits and costs	Economics: Definition and measurement of water related values	Making social adjustments to accommodate full and fair valuation of resources
Require the participation of all units of government and stakeholders in decision-making through a process of coordination and conflict resolution	Political science and sociology: knowing how to involve stakeholders meaningfully in a valid process	Committing to genuine stakeholder involvement
Promote water conservation, reuse, source protection, and supply development to enhance water quality and quantity	Environmental science and engineering: definitions of water management concepts	Deciding to accept these concepts into planning, design, and operation
Foster public health and safety	Public health and security: definition of water-related public health and security	Committing to investments and institutions to achieve them
Foster community goodwill	Sociology: understanding links between TWM and goodwill	Committing to social justice, equity, and fair treatment to all

see how knowledge and value questions leap out of it. Table 11-1 shows knowledge and value requirements, organized according to knowledge categories and arranged in the order of the TWM definition.

Summary points

- Valuing water is a big challenge to TWM and requires cultural change so that individuals value conservation of resources and practice sustainability in water use; organizations and their leaders see the entire TWM picture, learn why it is important, and commit to it in policy and action; and small dischargers and hydromodifiers understand their impacts and practice stewardship.
- Without stewardship, the cumulative effects of development will degrade water systems and the ecosystems that depend on them. This in turn will threaten life, health, and the environment. Stewardship is required because regulations and government programs are not enough by themselves.
- Barriers to stewardship include lack of knowledge, lack of incentives, and indifference.
- TWM emphasizes individual and corporate stewardship as an element of corporate social responsibility. AWWA supports corporate social responsibility and has partnered with an environmental organization to create an award for it.
- Environmental education is required to promote stewardship and is an important element of TWM. It offers opportunities for leadership to forge consensus about stewardship.

Review questions

1. Explain how the culture of the water industry affects how water is valued by customers. Explain what will be necessary to change this culture in the viewpoints of individuals and leaders, including those in large and small organizations.

2. What is meant by the *cumulative effects* of development? Will these degrade water systems and the ecosystems that depend on them? What are the consequences of any degradation that might occur?

3. If barriers to stewardship include lack of knowledge, lack of incentives, and indifference, identify which in your mind is the most important problem to be solved. Make suggestions for overcoming this barrier.

4. Explain what is meant by corporate social responsibility. Do you think it is a good concept or should business and utilities stick to their main missions without attending to social responsibilities? Justify your answers.

5. How can environmental education be used to promote leadership in forging consensus about stewardship?

References

Bok, Derek. 1996. *The State of the Nation: Government and the Quest for a Better Society.* Cambridge, Mass.: Harvard University Press.

Cone Inc. 2006. Multi-Year Study Finds 21% Increase in Americans Who Say Corporate Support of Social Issues Is Important in Building Trust. Corporate citizenship survey. http://www.coneinc.com/Pages/pr_30.html. Accessed November 15, 2006.

Covey, Stephen R. 1991. *Principle-Centered Leadership.* New York: Summit Books.

Henderson, David. 2002. *Misguided Virtue: False Notions of Corporate Social Responsibility.* London: Institute of Economic Affairs.

Manne, Henry G. 2006. Milton Friedman Was Right. *Wall Street Journal,* November 24.

North American Association for Environmental Education. 2007. http://naaee.org/.

Project WET. 2007. *About Us.* http://www.projectwet.org/aboutus.htm.

Richter, B. 2007. Blue Water Awards: Recognizing Utilities that Protect the Environment. *Jour. AWWA* 99 no. 4: 52 54.

Rolston III, Holmes. 1988. *Environmental Ethics: Values in and Duties to the Natural World.* Philadelphia: Temple University Press.

US Congress. 1990. *National Environmental Education Act of 1990.* PL 101 619.

CHAPTER 12

WATER INDUSTRY PROSPECTS AND POLICIES

In our growing world of more than six billion people, each sector of society has important roles in sustainable development. From headwaters to the oceans, the natural water environment depends on us for management and stewardship. Without these, development pressures will cause shortages, polluted water, degraded natural environments, and the loss of ecosystems.

Threats to the water industry

The environmental crisis is real, but it is perceived in different ways by different groups. To some groups, the crisis can be used to further their agendas. To others, it represents a problem to be confronted and solved. Regardless of how it is perceived, there are clear trends in some indicators, but we lack effective ways to measure overall environmental progress or decline, and we need a balanced report, not alarmist ones.

In the developed world, like the United States, Europe, Japan, and similar nations, the frameworks of water management practices are mostly established. Meanwhile, the global scorecard looks bleak because of growing nations either emerging from poverty or

> **From headwaters to oceans the water environment depends on us for management and stewardship**

trying to cope with it. Even in the United States, we see around us every day that myriad small water management actions create cumulative large impacts. Add to this the specter of global climate change, and the water industry has its hands full.

So the water industry and its leaders face fast-moving threats from global development and creeping threats like the frog in the pot of water being brought slowly to a boil. There is plenty of rhetoric about needed actions, but the question is how to translate our rhetoric into action. To do that, the water industry can take the lead in Total Water Management, not only to handle its own affairs but to also help government agencies, businesses, nongovernmental organizations, and citizen leaders see their important respective roles.

Total Water Management can address the threats through better water management reform and stewardship. Although the quest for a workable paradigm like TWM has been underway for at least 70 years, it is time now to embrace the ideas and overcome the barriers to implement it.

Where the water industry is heading

The water industry includes many organizations, from utilities and service agencies to government to private industry. You can view it like an iron triangle, as shown in Figure 12-1, where each corner of the triangle represents large stakeholder groups. Normally, an iron triangle represents government authorities, political stakeholder groups, and special interest groups. In this case, it has been adapted to show the tension between water supply organizations, government regulators, and industry support groups and businesses.

While the water industry cannot solve all water problems by itself, it can lead by marshaling its resources to show the way. It faces new challenges anyway to maintain and improve its basic services, and its customers expect it to take the lead in stewardship as well.

Water industry trends and issues

Pressure on water and natural systems will intensify as growth ratchets upward. The population of the United States has passed the 300 million mark and world population is heading toward seven billion, with a large percentage in the rapidly developing nations of China and India. Within the global village, billions of people seek to escape poverty and enjoy rising standards of living, which demand better resources management and delivery of public services.

With the Internet and increasing trade, globalization is affecting almost everyone. However, a safer and more cooperative world is an elusive

CHAPTER 12   WATER INDUSTRY PROSPECTS AND POLICIES

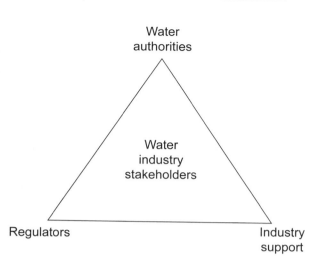

Figure 12-1. Water industry triangle

goal. Terrorism and natural disasters threaten water systems at a scale not known before. Regulatory affairs will remain a major concern of water managers. While the major environmental and health laws have already been enacted, new regulations continue to emerge, and the pressure on utilities to comply remains high. Infrastructure has emerged as the major concern among water utility managers, and the gamut of dams, water systems, and structures require monitoring, vigilance, and renewal.

New information technologies enable us to do a much better job in water management, if these can be embraced by utilities and their workforces. Modern information-based management systems are gaining more acceptance and are helping government to reinvent itself. Privatization and alternatives to government services remain in vogue, although expectations are now more realistic than a few years ago. Public involvement, including more emphasis on direct democracy, is more important than in the past.

In the United States, total water use remains steady, but demands are rising in water-short areas. Several forces combine to cause the cost of water to rise. Emphasis on full-cost pricing results from more attention to utility and enterprise management. Subsidies are reduced whenever possible.

While environmentalism is maturing, new emphasis on sustainability and worry about climate change have appeared on the radar screens of water utilities. Clearly, natural water systems cannot sustain much greater pressure, and there are limits on the availability of fresh water. Rising demands require more conservation and use of alternative sources.

Industry changes

As the water industry considers these trends, it is adapting itself to meet the challenges. The question is whether it will be reactive, or it will take the lead in solving emerging problems to move us toward sustainable development.

Of course, the water industry is a large and amorphous animal, but its core part, the water supply industry, has a clear vision of its future. AWWA's periodic survey of industry issues shows that the following five issues are at the top of the list of concerns among utility managers: maintaining infrastructure, complying with regulations, running the water business, obtaining and protecting source water supply, and building workforce capacity.

> Top water issues: infrastructure, regulations, business of water, source water supply, and workforce capacity

In addition, the industry is keeping its eye on other key issues, including security, macro factors (global warming, natural disasters, environmental activism, population growth), consumers, industry leadership, technology, energy, and wastewater (Runge and Mann, 2006).

State of the practice of TWM

TWM offers visionary concepts for the water industry, and all of them have not yet been implemented. However, TWM concepts offer powerful tools to lead us toward sustainability. Looking at its elements, you see that if it is successful, TWM offers hope for solving many of our most urgent water problems:

- Watershed plans that emphasize sustainability
- Economic, environmental, and social goals for water management
- Clear processes for planning, decision-making, monitoring, and adaptation
- Systems to allocate water efficiently among competing uses
- Defined roles and relationships
- Shared governance and coordination mechanisms
- Defined rules for consensus and conflict resolution
- Effective incentives for positive actions
- Transparency and accountability in planning and decision-making
- TBL reporting for unit and shared planning
- Effective assessment tools

People will disagree on this scorecard for practicing TWM, but Table 12-1 outlines my own assessment of how the United States and the rest of the world are faring in their attention to these elements.

This scorecard looks bleak, especially in the global category, and even in the United States many areas need much improvement. Areas needing the most attention in this nation are stewardship, equity, and conservation. Interestingly, these require three different types of responses: ethics, justice, and incentives, respectively. Ethics will address stewardship, justice aims to improve the lot of the poor and disenfranchised groups, and incentives will encourage all of us to do better. Our roles and responsibilities to promote these are discussed next.

Roles and responsibilities

The only way for the scorecard to improve is through an effective and sustained response to the challenges we face. The United States can exercise global influence, and at home we can improve our own situation a great deal. As Table 12-2 shows, each sector has clear roles and responsibilities. The roles column could almost form mission statements for utilities. For example, a wastewater utility could have a mission to

Table 12-1. Assessment scorecard by TWM element

TWM element	US score	Global score
Exercises stewardship for the greatest good of society and the environment	Mixed	Poor
Requires participation of all units of government and stakeholders	Good	Poor
Encourages planning and management in a dynamic process that adapts to changing conditions and local and regional variations	Good	Poor
Balances competing uses through efficient allocation	Fair	Poor
Conducts decision-making through process of coordination and conflict resolution	Good	Poor
Promotes water conservation, reuse, source protection, and supply development to enhance water quality and quantity	Fair	Poor
Addresses social values, cost-effectiveness, environmental benefits and costs	Good	Poor
Fosters public health, safety, community goodwill	Good	Poor
Exercise of citizen and corporate stewardship of water and related land resources*	Fair to poor	Poor
Complies with all laws and regulations*	Fair	Poor

* These elements were added to the original AwwaRF definition

"practice excellence in management of wastewater while complying with all regulations, going beyond the minimum to protect ecosystems, and reaching out to cooperate and to promote environmental education and social justice." That is a long statement but it captures the essence of what a wastewater utility is about, and the words can be crafted to suit the leaders of specific organizations.

The TWM tools look a lot like other management tools. To support TWM, they would be applied in different ways by the various stakeholders. For example, the strategic plans of water utilities would address how they support sustainability in taking water from the watershed, while those of wastewater utilities would address the discharge of contaminants to the environment.

Institutional arrangements

How barriers to TWM are institutional rather than technical was explained in chapter 10. Financial and social incentives drive land development, utilities are stressed and face limits in cooperating and taking on wider TWM roles, and intergovernmental coordination is a challenge. Meanwhile, the Tragedy of the Commons explains how water issues fall between the responsibilities of organized units and individuals.

Raising public understanding of the value of water is essential if environmental ethics are to become part of the social fabric of the nation. TBL reporting can raise public awareness, but who will do it for regional pictures that report the myriad of dispersed actions that escape the oversight of utilities and formal authorities?

Whether the issue is about large diversions and discharges or small actions, people do not follow sustainable practices unless regulatory controls or incentives direct them. The incentives of business and government have not yet led to the right mix of these controls and incentives, and many of them may work against sustainability. Can TWM bridge the gap, or is it doomed to remain a utopian concept that does not fit into the real world? What incentives would it take to transform TWM into a powerful force? Here are a few ideas:

- Control of the nation's water utilities is in their boards and elected leaders. If these are convinced about their roles to stimulate stewardship, we will see a big improvement.
- Opinion leaders and the media will not report about water concerns unless the public wants to read and hear about them. They can raise public awareness if they report more on water and environmental issues. It is up to these leaders to figure out how to make these issues compelling and interesting.

Table 12-2. Roles and TWM tools by sector

Sector	Roles	TWM tools
Water supply and wastewater utilities	Perform core missions Comply with regulations Protect ecosystems Cooperate in shared governance Offer environmental education Promote social justice Do TBL reports	Strategic plans Regulatory understanding Environmental programs Joint plans Environmental curricula CSR programs TBL report
Stormwater and flood control agencies	Perform core missions Comply with regulations Protect ecosystems Cooperate in shared governance Offer environmental education Do TBL reports	Strategic plans Regulatory understanding Environmental programs Joint plans Environmental curricula TBL report
Instream flow users (energy, navigation, recreation, fish and wildlife)	Comply with regulations Protect ecosystems Cooperate in watershed planning	Regulatory understanding Environmental programs Joint plans
Land developers, farmers, resource extractors, property managers	Comply with regulations Protect ecosystems	Regulatory understanding Environmental programs
Road departments	Comply with regulations Protect ecosystems Offer environmental education	Regulatory understanding Environmental programs Environmental curricula
Regulators	Perform core mission Reach beyond mission to promote stewardship Offer environmental education	Strategic plans Environmental programs Environmental curricula
Government	Perform policy and oversight roles, provide financial support	Policy analysis Assessment programs
Consultants, vendors, knowledge sector	Offer environmental education, develop products and services	Environmental curricula Opportunity to participate with water industry

- By the same token, political leaders can elevate water and environmental policies to a higher level of importance and work to create policies and metrics to assess progress in sustainable development. TBL reporting can go a long way toward raising awareness and support for these policies.
- Government agencies and regulators can move away from sole reliance on command-and-control measures and toward workable incentives for sustainable practices.

- Government can create incentives for business leaders to use their corporate social responsibility programs to focus on water and environmental issues. Business leaders can promote positive government action to create these incentives.
- Government, business, and citizen leaders can enable the knowledge industry to explain water issues and be scientific watchdogs of sustainable development.

Final word

TWM must work within our systems of government, incentives, and culture. It should be based on stewardship, but regulatory controls will still be required. After all, it is not a perfect world. There are roles for all sectors, but the water industry has special responsibilities.

As they deliver effective water services, water utilities can lead in promoting stewardship. They must help the water industry to beat the "it's not my problem" syndrome by going beyond their narrow interests. Opinion leaders and the media can explain TWM and publish report cards. They can serve as the conscience of TWM to identify and publicize justice issues that need attention. Political leaders can create policies that level the playing field and remove constraints faced by TWM practitioners. Government and regulators can translate policies that offer workable incentives for improvement. Business leaders can lead by complying with regulations and going the extra mile in CSR. The knowledge sector can work with citizen leaders to explain water issues and be scientific watchdogs.

Water utilities can issue TBL reports that highlight their corporate social responsibilities and work with others toward regional TBL reporting to promote coordination and shared approaches.

The emotion factor can sustain interest in TWM and its goals. People get interested in water issues, then lose interest because they last a long time. People must get excited about TBL report cards, stories of successes, and environmental improvement.

Education can promote the equity, justice, and incentives that are needed to improve TWM. Without it, it will not be possible to address the myriad small actions and hydrologic modifications that plague us.

Large gaps exist between national policies and local implementation. Devolving authority to local leaders, combined with more emphasis on stewardship, is more consistent with TWM than centralized command and control is.

Although climate change threatens us on a large scale, we can have water management that promotes sustainable development. Think about how the world will be if it works well. It's not a return to the Garden of Eden, but it is better than today.

TWM can pay substantial rewards. Its benefits will go beyond sustainable development to include positive impacts on environmental and science education; increasing respect for science and better governance; better utilities and public services; and the promotion of careers and business opportunities.

In the final analysis, TWM is about water industry leadership to achieve the triple bottom line goals that society expects from it. If the spirit and the letter of TWM are implemented, the result will be affordable and safe water that is provided to meet social needs without degrading the environment.

Summary points

- In the wealthier countries, the frameworks of water management practices are mostly established, but myriad small water actions create cumulative large impacts. In developing nations, TWM is more difficult because they are trying to overcome poverty and related political problems.
- The water industry can take the lead in Total Water Management through its own actions and by helping others to see their roles.
- Global trends mean that more effort is needed to mitigate pressure on natural water systems. Globalization affects everyone, and achieving a safer and more cooperative world is difficult. Security against terrorism and natural disasters is a more important goal than in the past. New regulations continue to emerge and the pressure on utilities to comply remains high. Along with security, regulations, and business matters, infrastructure has emerged as a major concern among water utility managers.
- TWM offers powerful tools for sustainability and offers hope for solving urgent water problems. Institutional barriers to TWM must be addressed if it is to be successful. These include financial and social incentives, intergovernmental coordination, and addressing issues that fall between the responsibilities of organized units and individuals.
- If environmental ethics are to become part of the national social fabric, raising public understanding will be required. Both regulatory controls and incentives will be required, but the right mixture is difficult to achieve.
- If water boards and leaders are convinced about stewardship, large improvements can result. Opinion leaders and the media should report more on water and environmental issues, even when they are not at a

crisis point. Political leaders can elevate water policies and create policies and metrics to assess progress in sustainable development.

Review questions

1. What are the most urgent water problems of developing countries? What reforms are most important in addressing them?

2. In wealthier countries, myriad small water actions create cumulative large impacts. Can these be controlled without infringing on individual liberty? If so, how?

3. Of the global economic, political, and environmental trends, which might have the greatest potential consequences for TWM?

4. Is security against terrorism and natural disasters a valid concern for TWM?

5. Do you hold out hope to increase public understanding and acceptance of environmental ethics? Why or why not?

6. If the water industry can take the lead in Total Water Management, how should its water boards and leaders respond to the challenges of leadership? What should be the roles of opinion leaders, the media, and political leaders?

Reference

Runge, J., and J. Mann. 2006. State of the Industry Report: 2006. *Jour. AWWA* 98 no. 10(October): 64–71.

APPENDIX A

AWWA AND AWWARF STATEMENTS ABOUT TOTAL WATER MANAGEMENT AND RELATED CONCEPTS

This appendix is presented to explain the origins and development of the conceptual framework of TWM. It presents four documents or statements about TWM from AWWA's or AwwaRF's archives or publications. It does not purport to be a complete record but will document at least part of the evolution of the TWM idea.

AWWA Policy Statement on Developing and Managing Water Resources

Adopted by the Board of Directors June 8, 1975, revised Jan. 31, 1982, Jan. 28, 1990, June 11, 2000, June 13, 2004

The American Water Works Association (AWWA) supports and promotes sound water resources planning and management which provides for an adequate supply of high-quality water for people. These efforts should give careful consideration to regional water resource conditions, environmental impacts, and project cost.

This must include the wise use of available resources, conservation of water by all practicable means, the reduction of pollution using best management practices, effective treatment and distribution of water, the encouragement of effective water reclamation and reuse when economically and technologically feasible, consideration of in-stream flow needs, and the taking of appropriate steps to protect life, property, and land from destructive forces of water.

Because comprehensive planning is a dynamic process, continual appraisal becomes the basis for the evolution of policies. It is equally important that the environmental implications of the plans be thoroughly considered in order that any adverse environmental impact be minimized.

It is with this background that AWWA sets forth the following principles by which the water supply profession can best meet its responsibilities to the public.

1. Where competition among water users occurs, high priority should be given to meeting human needs. To the maximum extent possible, higher quality water should be assigned to domestic use.
2. Each water source should be developed and managed with careful attention to the hydrologic and ecologic systems of which the particular source is a part. Surface and groundwater sources should be managed conjunctively.
3. The growing value of alternative water sources, such as desalted sea or inland saline water as public and industrial water sources, must be recognized. Such sources should be utilized where freshwater supplies are unavailable or inadequate, or where such converted waters are economically advantageous.
4. The responsible use of reclaimed water in lieu of potable water is encouraged for nonpotable uses. AWWA urges continued research to improve treatment technology, monitoring techniques, and the development of health-based drinking water standards, thereby assuring the safe use of reclaimed water.
5. The degradation of the quality of water supply sources has damaging effects on health, welfare, the economy, and the environment. Public water supplies, as an essential factor in the economy, are entitled to a good-quality source water.
6. Water is a renewable natural resource. It must be managed to best meet many needs. Every effective means to prevent and minimize waste and promote wise use should be employed by all entities, public and private, engaged in water resource activities.
7. Hydrologic, environmental, and other basic data are crucial to water resources development and management. Federal water resources data acquisition programs should be designed and conducted with attention to the full range of current and future uses by all entities, public and private. National databases on streamflow, groundwater levels, water quality, pollution threats, and land use should be made

easily available to all water suppliers for their use in water resources development and management.

The role of the federal governments in water resource programs and projects should be supportive and cooperative, not preemptive. Federal governments should recognize and respect the right of each state or province to control the use of its water and associated land resources, provided that management of the resources is responsible to clearly defined national and international needs. Regulations should not necessarily be uniform but should be tailored to regional circumstances and requirements.

AWWA White Paper on Total Water Management

The following is the text of the White Paper as published by AWWA (1994) in *MainStream*:

This paper offers the recommendations and rationale of AWWA for the application of total water management by water utilities and their regulators.

Principles of total water management outlined

The AWWA Executive Committee has approved a white paper on total water management.

This white paper is published to elicit discussion and consensus on issues of concern to the drinking water industry. The white paper and ensuing discussions will be used as the basis for AWWA's government affairs actions and public affairs programs.

Background

Regional, state, provincial, and local agencies face increasing frustrations as they attempt to plan for future community needs and implement their water supply, water quality, and wastewater management responsibilities. Environmental awareness, multiple laws, conflicting jurisdictions, scarce resources, increasing competition for available public funds, and increasingly factious citizen activism make their work appear impossible.

AWWA has endorsed the long-term goal of total water management, which is an attempt by the water supply industry to assure that water resources are management for the greatest good of people and the environment and that all segments of society have a voice in this process.

Today, environmental issues are being framed in terms of watershed management by federal agencies. The US Geological Survey has identified 21 major watershed basins, and each state or province is further divided into smaller watersheds that feed the major drainage systems. President Clinton's Clean Water Initiative, submitted to Congress on Feb. 1, 1994, supports a new provision in the Clean Water Act to establish statewide programs for comprehensive watershed management.

Total water management

Total water management recognizes the paradigm shift from considering water available in unlimited quantities to understanding water supply as a limited resource.

All water issues revolve around three factors: water quantity, water quality, and establishing priorities to deal with the limitations of water quality and quantity. The need to prioritize is being debated at the national level, accented by conflicting uses. Recent allocation of waters in the Pacific Northwest for fisheries and Native Americans and the reallocation of water from the Edwards Aquifer in Texas for endangered species bear this out.

The major challenge to the drinking water industry is developing the process to establish priorities. Water by its very nature is an integral part of every environmental issue and a basic need for the public welfare and prosperity. Thus the water cycle must be recognized in all forms in the environment—from ice to liquid to vapor.

Total water management should consider the integration of the complete water cycle. Legislation must give opportunities to consider and determine the interrelationships between all aspects of the environment and society on a regional basis rather than dealing with each issue discretely and within limited parameters.

The program must begin at the local level and integrate the activities of local, state, provincial, and federal governments if total water management programs are to succeed.

Stewardship

The water utility industry cannot be concerned only with providing potable water. The role of the utility in providing safe water for human uses must be expanded to include good stewardship. This effort requires water utilities to strive to not only be leaders but recognized as stewards of good water policy.

Land and water resource management must be integrated at the local level. Water utilities must position themselves to effect change in the way that land and water resources are currently managed. This could ultimately lead to changes in demand management and the identification of water reuse as a constraint for land use in water-short areas.

Government role in total water management

There is an urgent need for a unified water resources policy that observes the principles of integrated land and water resource planning and management under a watershed framework and is based on rational priorities. This would relieve the patchwork of conflicting objectives and jurisdictions at the federal, state, and local government levels, as well as address regional differences, urban and rural distinctions, competition between cities and agriculture for water, and interbasin transfers.

During the first half of this century, an extensive system of water storage was constructed for municipal supply, agricultural irrigation, and flood control. These facilities are the United States' most important water assets and form the backbone for the United States to structure a more effective total water management program. They must be better integrated to meet future water needs. Conservation of municipal and agricultural uses—combined with water reuse, reallocation of resources, and watershed management—will be necessary to meet the challenge of a national water program for sustainable development.

A new federal water policy must integrate planning, management, and development to protect surface water and groundwater resources under a watershed framework. It must be based on the principles of pollution prevention and resources conservation incorporated into a sustainable development strategy. The policy should also be designed to incorporate concern for water resources into every aspect of human activity. The policy should strive to integrate institutions, economics, ecology, and technology into a common objective. Furthermore, policy implementation should be delegated to the states, limiting the federal role primarily to technical assistance and interstate water management issues.

Watershed management

Watershed-based management on a subdrainage basis is one tool that can be implemented for the protection of water resources. Because most economic and natural events that affect the quality of water resources occur principally within watershed boundaries, watershed boundaries are the most sensible way of taking action to restore and protect water resources. This approach provides a framework that would supersede international political boundaries to evaluate and solve natural resource problems such as water quality.

The US Geological Survey's 21 major water-resource regions with their many subdivisions provide a framework for the establishment of a basis for watershed management in the United States. The Candiadian and Mexican equivalent, further divided by the state and provincial watersheds, should also establish a framework. These USGS hydrologic units, which encompass the drainage areas of the major river systems, provide the flexibility to address water quality problems at the appropriate level.

Water resource management

Water supports life—from the basic needs of living organisms to complex habitats and recreational and aesthetic environments, as well as public drinking water requirements. The water industry must consider the total interaction of water with the environment, including balancing human and ecological risk and the preservation and restoration of ecosystems. The challenge is in assuring public health, safety, and welfare—which must take precedence—while achieving this balance.

Water availability and allocation can be a constraint on development and economic options. For example, the Endangered Species Act can have an enormous impact on a local water utility because the act prevents the drawdown of an underground aquifer if it feeds streams critical to an endangered species impact. A similar or corresponding act should address the needs of society.

Water conservation

Water is a renewable but finite natural resource. Water conservation considerations should be a part of any utility's water resources planning. Conservation, encompassing supply and demand management, is appropriate to some degree for all utilities and not just those in water-short areas.

To convince the local population that water conservation makes good water and economic policy, however, local water utilities will need to educate consulters about the benefits of regionally appropriate conservation measures and resources planning. This may be a daunting task for those utilities in areas where water resources are plentiful.

Public support

Public support for total water management decisions is critical for the water manager. Water suppliers have a distinctly public role by virtue of contributing to the public health, as well as managing a sustainable natural resource. Utilities will play a major role in the process of disseminating information through a variety of forums. For issues that affect the community and its water resources, water utilities will play an important part in enlisting public participation in those decisions.

The water users, as well as the general public who may be affected by total water management decisions, should be a part of the decision-making process. The public should be included in analyzing alternatives, and evaluating relative-risk reduction and the economic effects of alternatives. Relative-risk reduction must include adequate regulatory flexibility so that environmental problems can be reevaluated from a risk reduction benefit and cost perspective. Remedies must be achieved through priorities set through public choices.

The public must have a voice in decisions of significant impact, such as water conservation or curtailment as a solution to water shortages during drought periods, balancing completing needs for the resource, and growth and economic development. These decisions will need to be made on a regional or even multistate basis.

Political support

Political leadership by local and national representatives will be required to achieve the goals of a total water management program, and AWWA asks the national political leadership to support the effort to accomplish our vision of total water management. The technical knowledge of AWWA is available and stands ready to assist governmental leaders in developing a national water policy that incorporates total water management.

AwwaRF definition of Total Water Management (1996)

The details of TWM were drawn out by a group of more than 30 water industry professionals at an AwwaRF workshop (1996). Here is the definition that was developed after two days of intensive work:

> Total Water Management is the exercise of stewardship of water resources for the greatest good of society and the environment. A basic principle of Total Water Management is that the supply is renewable, but limited, and should be managed on a sustainable use basis. Taking into consideration local and regional variations, Total Water Management: encourages planning and management on a natural water systems basis through a dynamic process that adapts to changing conditions; balances competing uses of water through efficient allocation that addresses social values, cost effectiveness, and environmental benefits and costs; requires the participation of all units of government and stakeholders in decision-making through a process of coordination and conflict resolution; promotes water conservation, reuse, source protection, and supply development to enhance water quality and quantity; and fosters public health, safety, and community goodwill.

AWWA definition of Total Water Management, from the *Drinking Water Dictionary* (2000)

The management of water resources with a comprehensive approach to balancing resources, demands, and environmental issues. Total water management considers water supply, water quality and treatment, storage, conveyance, potential use of alternative water supplies (such as water reuse or desalting of saline waters), conservation and demand-side management, effects of water users, and environmental needs and concerns. (Symons, Bradley, and Cleveland, 2000)

References

AWWA. 1994. Principles of Total Water Management Outlined. *MainStream* 38 no. 11 (November): 4, 6.

AWWA. 2004. Policy Statement on Developing and Managing Water Resources. Denver, Colo.

AwwaRF. 1996. *Total Water Management Workshop Summary*. Draft. Seattle, Wash., August 18–20. Denver, Colo.: Awwa Research Foundation.

Symons, J.M., L.C. Bradley, and T.C. Cleveland, 2000. *The Drinking Water Dictionary*. Denver, Colo.: American Water Works Association.

APPENDIX B

LIST OF ACRONYMS

APA	Administrative Procedures Act
ASCE	American Society of Civil Engineers
ASDSO	Association of State Dam Safety Officials
AWWA	American Water Works Association
AwwaRF	American Water Works Association Research Foundation
BCA	Benefit–cost analysis
bgd	Billion gallons per day
BMP	Best Management Practices
BSC	Balanced scorecard
BTWF	Big Thompson Water Forum
CAFOs	Concentrated animal feeding operations
CAFR	Comprehensive annual financial report
CCR	Consumer Confidence Report
CEQ	Council on Environmental Quality
cfs	Cubic feet per second
CSR	Corporate social responsibility
CWA	Clean Water Act
CWS	Community water system
DBP	Disinfection by-product

EIS	Environmental Impact Statement
EQ	Environmental Quality
ESA	Endangered Species Act
EU	European Union
EWRI	Environment and Water Resources Institute
FASB	Financial Accounting Standards Board
FEMA	Federal Emergency Management Agency
FERC	Federal Energy Regulatory Commission
FPA	Federal Power Act
GAO	Government Accountability Office
GASB	Government Accounting Standards Board
gpcd	Gallons per capita per day
IBT	Interbasin transfer
IFIM	Instream Flow Incremental Methodology
IJC	International Joint Commission
IPART	Independent Pricing and Regulatory Tribunal (Australia)
IRP	Integrated resource planning
IRS	Internal Revenue Service
ISF	Instream flows
IWRM	Integrated Water Resources Management
LID	Low-impact development
M&I	Municipal and industrial
MCDA	Multicriteria decision analysis
MCL	Maximum contaminant level
MCLG	Maximum contaminant level goal
mgd	Million gallons per day
MUA	Multiattribute analysis
NCWS	Noncommunity water system
NED	National Economic Development
NEPA	National Environmental Policy Act
NGO	Nongovernmental organization
NPDES	National Pollutant Discharge Elimination System
NPDWR	National Primary Drinking Water Regulations
NPS	Nonpoint sources
NRDC	Natural Resources Defense Council

O&M	Operations and maintenance
P&G	Principles and Guidelines
P&S	Principles and Standards
PCB	Polychlorinated biphenol
PHABSIM	Physical Habitat Simulation
PUC	Public utility commission
PVO	Private volunteer organization
RBC	River basin commission
RCRA	Resource Conservation and Recovery Act
RD	Regional development
SDWA	Safe Drinking Water Act
SEC	Securities and Exchange Commission
SEPA	State Environmental Policy Acts
SIA	Social impact assessment
SMARTT	Source Management and Rotation Technology Tool
SPU	Seattle Public Utilities
SWB	Social well-being
SWFC	Stormwater and flood control
SWFWMD	South Florida Water Management District
TBL	Triple Bottom Line
TG	Thousand gallons
THM	Trihalomethanes
TMDL	Total maximum daily load
TWM	Total Water Management
UIC	Underground injection control
UN	United Nations
UNESCO	United Nations Education, Science, and Cultural Organization
USEPA	United States Environmental Protection Agency
USDI	United States Department of the Interior
USFWS	United States Fish and Wildlife Service
USGS	United States Geological Survey
WASH	Water and Sanitation for Health
WET	Water Education for Teachers
WFD	Water Framework Directive
WQ 2000	Water Quality 2000

WRDA	Water Resources Development Act
WRPA	Water Resources Planning Act
WWAP	World Water Assessment Programme
WWTP	Wastewater treatment plant
7Q10	Seven-day ten-year low flow

INDEX

Note: *f.* indicates a figure; *n.* indicates a (foot)note; *t.* indicates a table.

A

Accountability, 87
Accounting stance, 146–147
Acronyms, 283–286
Albemarle-Pamlico estuary program, 82
Albertson, Maurice, 98*n.*
American Society of Civil Engineers
    Environment and Water Resources
        Institute (EWRI), 98
    Water Resources Planning and
        Management Division, 98, 98*n.*
American Water Works Association (AWWA)
    Blue Water Award, 256
    and performance indicators, 124
    policy statement on developing and
        managing water resources (text),
        275–277
    regionalization, defined, 109
    and TWM, 4
    and water education, 259
    and water resources planning, 98–99
    White Paper on TWM (text), 277–281
Appropriation doctrine, 224, 226
Assessment, 85–86, 85*f.*
    of source water quality by EPA, 35–36, 36*t.*
AwwaRF
    author's expansion on definition of TWM, 62–63
    on current status and benefits of TBL reporting, 125–126
    definition of TWM, 56, 56*t.*
    detailed definition of TWM (text), 58–59, 281

Managing Constraints to Water Source Development, 242
The Value of Water: Concepts, Estimates, and Applications for Water Managers, 149

B

Balanced scorecard, 119–121, 120*f.*
Basin plans. *See* Watershed management
Benefit–cost analysis, 142, 146, 148–150, 158
    environmental, 179–181
Benefits, defined, 145, 149
Best management practices (BMPs), 73
    incentives, 82–84
Big Thompson Water Forum (Colorado), 75, 86
Blue Water Award, 256
BMPs. *See* Best management practices
Bok, Derek, 191–192, 259
Boundary Waters Treaty of 1909, 75

C

Case studies, use of, 13–14
Chesapeake Bay Program, 87
China, 266
Chowan River Restoration Project, 82
A Civil Action, 196
Clean Water Act, 35, 72, 169, 210, 211*f.*, 214–215, 231
    and Corps of Engineers, 81
    on environmental monitoring and assessment, 181–182

and multiobjective (TBL) reporting, 127
  programs of, 216t.–217t.
  and public health and safety, 196
  on stormwater quality, 228
  and supply allocation, 79
  and Total Maximum Daily Load (TMDL)
    program, 180–181
Coastal waters, 82
  environmental issues, 169
  See also Estuaries
Colorado Watershed Assembly, 183
*Community-Based Watershed Management
  Handbook*, 71
Competing uses, 77–78, 188
Consensus
  as goal in water allocation, 79
  levels of, 108
Conservation ethic, 83
Conservation. *See* Water conservation
Consumer Confidence Reports (CCRs), 127
Contaminants, 197–198
  list of, and potential effects, 199t.
Coordination
  functional, 109
  intergovernmental, 108–109
  of knowledge areas and disciplines, 109
  mechanisms, 75–77, 76f., 77, 78f.
  transboundary issues, 77
Corporate social responsibility, 13, 23, 24f.,
    255–256, 262
  balancing with utility responsibilities, 68,
    68f.
  implementing, 88–89
  and outward-looking agencies, 23, 24f.
  and reporting, 121
  See also Social issues; Stewardship
Cost, defined, 145, 149
Cost-effectiveness, 155
Council on Environmental Quality, 128
Cousteau, Jacques, 74
Cryptosporidium, 198
Customer service, as basis for TWM, 69

D

Dams
  large, 31–32
  ownership of, 32
Developing nations, 266
Diversions, 28, 29t., 33–34, 39
The Drinking Water Dictionary, TWM
  definition, 56, 56t., 282

E

Earth Day, 163, 256
Economics
  defined, 144
  distinguished from finance, 143–144,
    144f., 158
Education. *See* Environmental education

Eisenhower, Dwight, 251
Endangered Species Act, 81, 210, 211f.,
    219–220
Engergy Policy Act of 1992, 60
Environmental benefit–cost analysis, 179–180
  Tampa Bay's multicriteria decision analysis
    (MCDA) tool, 118–119, 180
  Total Maximum Daily Load (TMDL)
    program, 180–181
Environmental education, 256–258, 262
  and business, 258, 260
  and government, 258, 259
  knowledge and value requirements,
    260–261, 261t.
  and leadership, 260
  and media, 258–259
  multiple-sector involvement in, 258
  and nongovernmental organizations
    (NGOs), 259
  and private volunteer organizations
    (PVOs), 259
  and schools, 258
Environmental ethics, 13, 256, 273
  knowledge and value requirements,
    260–261, 261t.
  See also Stewardship
Environmental impact statements, 72, 181
Environmental issues, 22–23, 39
  coastal water vulnerability, 169
  deforestation, 168
  differing views of, 171
  discarded prescription drugs, 170–171
  environmental water needs, 165–166, 186
  global climate change, 166, 168, 171
  habitat loss, 168, 174
  international reports on, 167
  and lakes, 169–170
  population growth, 162, 162f., 198, 267,
    268
  species adaptability and diversity, 169, 171
  species at risk, 167
  state of the environment, 163–167, 186
  stream impacts and pollution, 168–169,
    171
  urbanization and land development, 168
  USEPA report on, 166–167, 168t.
  water quality impacts listed by source, 169,
    170t.
  water scarcity, 165, 186
  See also Natural water systems
Environmental monitoring and assessment,
    181–182, 181f., 188
Environmental quality, 102, 125
Environmental reports, 128
EPA. *See* US Environmental Protection
  Agency
Erin Brockovich, 186
Estuaries
  EPA water quality assessments, 35–36,
    36t.
  management challenges, 82
  National Estuary Program, 70–71

pressures and problems, 176–177
Targeted Watershed program (USEPA), 186, 187t.
See also Coastal waters
European Union. *See* Water Framework Directive

F

Federal Emergency Management Agency (FEMA), 221
Federal Energy Regulatory Commission (FERC), 220, 221
Federal Power Act (FPA), 79, 210, 211f., 220–221
Field Guide to SDWA Regulations, 215
Finance
  defined, 144
  distinguished from economics, 143–144, 144f., 158
Financial Accounting Standards Board (FASB), 121
Flood Control Act of 1917, 59
Flood Control Act of 1936, 146
Flood Disaster Protection Act, 221
Florida Panhandle mutual aid agreement, 69
Fort Collins (Colorado) metering case study, 84
Framework, defined, 56, 57t.
Friedman, Milton, 256

G

Gap analysis, 246, 248, 249t.–250t.
General Motors quote ("what's good for America"), 5, 5n.
Global climate change, 166, 168, 171, 267, 268
Global Reporting Initiative, 117
Global Water Partnership, 60
Globalization, 266–267, 273
Governance, defined, 103, 111
Government
  AWWA on government and political role in TWM, 278–279, 281
  and encouragement of TWM, 271–272, 273
  and environmental education, 258, 259
  levels of, and laws, 212, 212t.
Government Accountability Office (GAO), 121
Governmental Accounting Standards Board (GASB), 124
Great Lakes Regional Collaboration of National Significance, 75
Groundwater, 176

H

Hardin, Garrett, 23
Hierarchy of human needs, 193–194, 194f.
Hobbes, Thomas, 195

Hunt, James, 82
Hydrologic modifications, 28, 29t., 39, 173
  defined, 36–37
  identifying, 37–38
  types of, 37

I

India, 266
Information technologies, 267
Institutional issues, 235–238, 236f., 237t., 250–251
  authorities, 239, 251
  culture of the water industry, 240, 251
  dealing with complexity, 251
  gap analysis, 246, 248, 249t.–250t.
  incentives, 239, 240t., 251
  industry constraints, 242–243
  institutional analysis, 245–246, 248
  law vs. reality, 237–238
  laws, 239, 251
  nonpoint source pollution, 243–245
  obstacles to TWM, 240, 241t.–242t.
  political model of water resources planning, 246–248, 247f.
  regulations, 239, 251
  roles and relationships, 240, 248, 250t., 251
  water industry stakeholders (iron triangle), 238, 251
Instream flow, 173–174
Instream Flow Incremental Methodology (IFIM), 174
Integrated Resource Planning (IRP), 60, 61, 98–99
Integrated Water Resources Management (IWRM), 4
  comparison with TWM, 59–61, 89
  defined, 60
  evolution of, 59, 59f.
  policy sectors and purposes, 59, 59f.
Internal Revenue Service (IRS), 121
International Drinking Water Supply and Sanitation Decade (1980s), 198
International Joint Commission, 75
"Invisible hand," 10, 10n.
Iron triangle, 238, 251, 266, 267f.
Irrigation, 151–152
IWRM. *See* Integrated Water Resources Management

J

Journal AWWA, 215

K

Koelzer, Victor, 98n.

L

Lakes

as elements of natural water systems, 174–175
environmental issues, 169–170
EPA water quality assessment, 35–36, 36t.
Laws, 207–208
and administrative systems (water allocation), 223
appropriation doctrine, 224, 226
Clean Water Act, 214–215, 216t.–217t.
common enemy rule, 227
distinguished from regulations, 211–212, 228, 231
effect on management choices, 210, 211f.
Endangered Species Act (ESA), 219–220
federal, 214–222, 214n.
Federal Power Act (FPA), 220–221
growth of water laws, 207, 208f.
impingements on stewardship, 208–210, 231
on instream flow, 225–226
on interbasin transfer, 226
international, 230
interstate compacts, 222
legal categories that include water law, 212, 212t.
and levels of government, 212, 212t.
local, 227–228, 231
management tasks and related legal framework, 213, 213t.
National Environmental Policy Act (NEPA), 219
National Flood Insurance Act, 221
natural flow rule, 227
and public utility commissions, 226
reasonable use doctrine, 223, 227–228
riparian doctrine, 223
role of courts, 230, 231
Safe Drinking Water Act, 215–216
state, 222–226, 231
State Environmental Policy Acts, 222
stormwater ordinances, 227–228
on surface water allocation, 222–224
treaties, 222
Water Resources Development Acts, 221–222
water use restrictions, 227
See also Regulations
Locke, John, 195
Longs Peak Working Group, 78, 165–166, 165n.

M

Managing Constraints to Water Source Development, 242
Maslow, Abraham, 193
Maximum contaminant levels, 217, 218
Meters and metering, 84
Multiattribute analysis (MUA), 68
Multicriteria decision analysis (MCDA), 118–119, 133–134
Multiple barrier approach, 217

N

National economic development (NED), 102, 125, 148
National Environmental Education Act, 257
National Environmental Policy Act (NEPA) of 1970, 72, 128, 181, 219
and environmental education, 256
National Estuary Program, 70–71, 177
National Flood Insurance Act, 221
National Hydrography Dataset, 37
National Marine Fisheries Service, 219
National Pollutant Discharge Elimination System (NPDES) permits, 34, 210, 211f.
National Primary Drinking Water Regulations (NPDWRs), 217–218
National Regulatory Research Institute, 226
National Resources Planning Board, 97
National Water Assessments, 127
National Watershed Forum (2001), 183–184
Natural disasters, 221, 267, 268, 273
Natural flows, 172, 172n.
Natural Resources Defense Council, 21
Natural water systems, 171–172
defined, 172
estuary functions, 176–177
groundwater, 176
lakes and reservoirs, 174–175
rivers and streams, 173–174
scope of, 172
watersheds, 172–173
wetlands, 175–176
Nature Conservancy, 174
Blue Water Award, 256
New Deal era, 74, 97
Nonpoint source discharges, 28, 29t., 35, 39
and stream impairment, 35–36, 36t.
Water Quality 2000 on related institutional issues, 243–245
North American Association for Environmental Education, 257

O

Opportunity cost, 143, 156

P

Paradigm, defined, 56, 57t.
Physical Habitat Simulation System (PHABSIM), 174
Planning. *See* Water resources planning
Play. *See* "Vienne River" case study
Point source discharges, 28, 29t., 34–35, 39
and stream impairment, 35–36, 36t.
Policy development, 65–66
regulatory- vs. market-based approaches, 66
Ponds, 32
Population growth, 162, 162f., 198, 267, 268
Postel, Sandra, 1

Practices
  defined, 89
  See also Best management practices
Price, defined, 145
Principles
  assessment and TBL reporting, 85–86, 85f.
  commitment to coordination, 75–77, 76f.
  corporate social responsibility programs, 88–89
  customer service first, 69
  defined, 56, 57t., 89
  effective policies, 65–66
  effective TWM process, 71–72
  efficient, equitable water resource allocation, 77–79
  incentives for conservation and BMPs, 82–84
  list of, 89–90
  regulatory effectiveness, 80–82
  roles and relationships, 68–69, 68f., 240, 248, 250t.
  shared goals, 70–71
  shared governance, 67–68
  transparency and accountability, 87
  watershed basis for planning, 72–75
  workforce and public participation in stewardship, 87–88
Processes
  and adaptation, 71
  defined, 56, 57t., 89
  and environmental impact statements, 72
  responsibility for establishing, 71–72
Project WET (Water Education for Teachers), 257
Public health and safety, 196–200, 204
  USEPA list of contaminants and potential effects, 198, 199t., 200f.
Public involvement, 267
  AWWA on, 280–281

Q

R

Reagan, Ronald, 97
Regional development (RD), 102, 125, 148
Regionalization, 109–110
  defined, 109
  and TBL reporting, 129–131
  See also Watershed management
Regulations, 11–12, 207–208, 228–229, 267, 268, 273
  agencies, 81
  as coordinating mechanisms between business and environment, 228–229
  distinguished from laws, 211–212, 228, 231
  enforcement of, 80–81, 229
  and estuary management, 82
  impingements on stewardship, 208–210, 231

vs. market-based approaches, 66
  need for, 80
  stewardship beyond regulations, 12
  See also Laws
Reservoirs, 32
  as part of natural water systems, 174–175
Resource Conservation and Recovery Act (RCRA), 218
Riparian doctrine, 223
Rivers and streams, 173
  EPA water quality assessment, 35–36, 36t.
  and hydrologic alteration, 173
  instream flow, 173–174
  See also "Vienne River" case study; Watershed management
Roles and relationships, 68–69, 106, 107t., 240, 248, 250t. *See also* Shared governance; Water resources planning
Rotary Club, 259
Rousseau, Jean-Jacques, 195

S

Safe Drinking Water Act, 210, 211, 211f., 215, 231
  and chemical problems, 204
  and Consumer Confidence Reports (CCRs), 127
  and maximum contaminant levels, 217, 218
  and multiple barrier approach, 217
  and National Primary Drinking Water Regulations (NPDWRs), 217–218
  primary and secondary standards, 217, 218
  and public health and safety, 196
  state responsibilities, 217, 218
  and USEPA, 217, 218
San Antonio (Texas) River Walk, 202
Schad, Ted, 98n.
Seattle (Washington) Public Utilities, TBL reporting case study, 132, 133t.
Securities and Exchange Commission (SEC), 121
Senate Select Committee on Water Resources, 97, 98n.
Shared goals, 70–71
Shared governance, 67–68, 93, 103–105, 110, 110f., 111
  coordination task, 106, 112
  decision problems requiring, 105–106
  governance, defined, 103, 111
  integration task 106, 112
  intergovernmental coordination, 108–109
  and planning, 93–96
  regionalization, 109–110
  See also Roles and relationships; Water resources planning
Smith, Adam, 10n.
Snow, John, 196
Social capital, 254
Social equity analysis. *See* Social impact assessment

Social impact assessment (SIA), 154, 202
Social issues, 191–193
  community goodwill, 201–202
  culture, recreation, and other social effects, 202
  equal opportunity, 201, 204
  hierarchy of human needs, 193–194, 194f.
  indicators, 194–195
  public health and safety, 196–200, 199t., 200f., 204
  rights and responsibilities, 203
  safety and security, 200–201, 204, 268
  and social contract, 195, 203–204
  TWM and broad societal involvement, 9–10, 11f., 19, 26–28, 39
  TWM and mechanisms to advance society, 195–196, 197t.
  and value of water, 144, 146–147, 147t.
  and water allocation, 141–143, 142f.
  and water industry outreach, 191, 192f.
  See also Corporate social responsibility
Social welfare function, 125
Social well-being, 102, 125, 193
Source Management and Rotation Technology Tool (SMARTT), 68, 180
Southwest Florida Water Management District, 67
State Environmental Policy Acts, 222
The State of the Nation, 191–192
Stewardship, 253–254, 262
  AWWA on, 278
  and corporate social responsibility, 13, 23, 24f., 255–256, 262
  going beyond regulations, 12, 262
  individual, 254–255
  as key point of TWM, 27
  and laws and regulations, 208–210, 231, 239
  as public responsibility, 12, 87–88
  as shared responsibility, 27–28
  and utility workforce, 87–88
  water utilities' role in promoting, 272–273
  See also Corporate social responsibility; Environmental education; Environmental ethics
Strategic planning, 246
Streams. *See* Rivers and streams
Surface Water Treatment Rule, 211
Sustainable development, 3–4, 17, 161, 186
  balance point, 7, 7f.
  barriers to, 9–10
  and natural systems, 162–163
  and shared responsibility, 26–28, 39
  and TBL reporting, 117, 134
  threats to, 28, 28f., 29t.
  watershed basis for planning, 72–75, 107–108
  See also Triple Bottom Line
Sydney (Australia) Water, TBL reporting case study, 131–132
Systems thinking, 246

T

Tampa Bay (Florida) Water
  and shared governance, 67–68
  use of multicriteria decision analysis (MCDA), 118–119, 180
Targeted Watershed program (USEPA), 186, 187t.
TBL. *See* Triple Bottom Line
Terrorism, 267, 273
Tocqueville, Alexis de, 258
Total Maximum Daily Load (TMDL) program, 180–181
Total Water Management, 1–2
  assessing environmental benefits and costs, 179–181
  AWWA White Paper on principles of, 277–281
  balance among environmentalists, water managers, and customers, 7–8, 8f., 17
  as balancing act, 2, 2f., 148f.
  balancing utility responsibilities and environmental considerations, 3–4, 5–7, 6f.
  and broad societal involvement, 9–10, 11f., 19, 26–28, 39
  characteristics of, 64
  comparison with IWRM, 59–61, 59f., 89
  comparison with Water Framework Directive (European Union), 61–62
  as comprehensive approach, 163–164, 165f.
  defined, 1, 24
  defined (AwwaRF), 27, 56, 56t., 58–59, 62–64, 281
  defined (*Drinking Water Dictionary*), 55–56, 56t., 282
  elements of, 15, 16f.
  and environmental leadership, 260
  evolution of, 59, 59f.
  framework, 2, 3t., 17, 56, 57f.
  and government, 271–272, 273
  implementing, 270–273
  knowledge and value requirements, 260–261, 261t.
  and laws and regulations, 11–12
  leadership role of utilities, 16
  and media, 270, 273
  obstacles, 240, 241t.–242t.
  principles, practices, and processes, 10–11, 56, 57f., 64, 64t.–65t., 65f., 89
  and public (political) responsibility, 12–13
  relation to large and small actions, 26, 27f.
  responsibilities for, 209, 209t.
  roles and responsibilities by sector, 269–270, 271t.
  state of the practice, 268–269, 269t.
  and stewardship beyond regulations, 12
  and utility boards and leaders, 270, 273
  and valuing water, 253–254, 262
  "Vienne River" case study (a play), 43–54

See also Institutional issues; Stewardship; Sustainable development; Triple Bottom Line; Water management
Tragedy of the Commons, 10, 11, 23–24, 39, 270
   assessment and TBL reporting as way to counter, 85, 85*f.*
Transparency, 87
Triple Bottom Line (TBL), 9, 9*n.*, 10, 70
Triple Bottom Line reporting, 85–86, 85*f.*, 115–116, 272
   attributes of indicators, 122–124, 123*t.*
   benefits of, 126
   business vs. government reporting, 121–122, 122*f.*
   compared with balanced scorecard approach, 119–121, 120*f.*
   compared with multicriteria decision analysis (MCDA), 118–119, 133–134
   condensing complex information, 123–124, 123*f.*
   current status, 125–126, 127–128
   developing indicators, 122, 128, 129*t.*
   displaying economic, environmental, and social impacts, 15–116, 116*f.*, 124
   environmental indicators, 124, 134
   goals and specific measures, 126–127, 126*t.*
   indicators, 121–124
   by individual utilities, 128, 129*t.*
   integrity in, 133
   issues to include, 128, 129*t.*
   knowing the audience, 116–117
   as multicriteria scorekeeping, 118
   need for clear indicators, 122, 124, 134
   regional, 128–131
   Seattle Public Utilities case study, 132, 133*t.*
   social indicators, 124, 134
   social welfare function, 125
   as sustainability reporting, 117, 134
   Sydney Water case study, 131–132
TWM. *See* Total Water Management
Two Forks Project (Colorado), 72

U

United Nations
   International Drinking Water Supply and Sanitation Decade (1980s), 198
   and international water disputes, 230
   UNESCO World Water Assessment Program, 167, 183
Urban areas
   environmental issues, 168
   and safe drinking water and sanitation, 198–199, 200*f.*, 204
   shantytowns, 198
US Agency for International Development, 199–200
US Army Corps of Engineers, 31, 81
   and dam building, 97

US Bureau of Reclamation, 31, 151–152
US Constitution, 12–13
US Department of the Interior, 219
US Environmental Protection Agency (USEPA)
   and Chowan River Restoration Project, 82
   Draft Report on the Environment (2003), 166–167, 168*t.*
   list of contaminants and potential effects, 198, 199*t.*
   and National Estuary Program, 70–71, 177
   and performance indicators, 124
   Reach File, 37–38
   on stormwater, 228
   Targeted Watershed program, 186
   water pollution data, 35–36, 36*t.*
US Fish and Wildlife Service, 81
   Instream Flow Incremental Methodology (IFIM), 174
   on species at risk, 167
US Forest Service, 81
US Geological Survey (USGS)
   mapping in identification of hydrologic modifications, 37
   National Water Summaries, 127
   US water-use statistics (2000), 34, 34*t.*
   water-resource regions and subdivisions, 279
   water-use database, 34
US Water Resources Council, 101, 112, 158, 203

V

Value, defined, 145
Value of water, 137–138
   agricultural, 151–152
   and balancing uses, 155
   and benefit–cost analysis, 148–150, 158
   concepts of, 145, 146*f.*
   and cost of service, 150–151
   and cost-effectiveness, 155
   in dilution of wastewater, 152–153
   economic and societal, 144
   economics vs. finance, 143–144, 158
   environmental, 154, 161
   fictitious market and auction example, 138–139, 139*f.*
   for fisheries, 153
   and flood control, 154
   for hydropower, 153
   individual accounting stance, 146, 147
   market value, 149, 150
   market vs. nonmarket value, 156
   municipal and industrial, 150–151
   for navigation, 153–154
   next least expensive option, 149–150
   as political issue, 140, 157
   organizational accounting stance, 146–147, 147*t.*, 157
   recreational, 153
   societal accounting stance, 146–147, 147*t.*

and TWM, 137–138, 157, 253–254, 262
and utility monopolies, 156–157
in water quality improvements, 152–153
See also Water allocation
The Value of Water: Concepts, Estimates, and Applications for Water Managers, 149
"Vienne River" case study (a play), 43–54
Virgin water systems, 172, 172n.

W

Water allocation
consensus as goal in, 79
coordination mechanisms, 78, 78f., 79
and economics, 141
efficiency and equitableness in, 77–79
and opportunity cost, 143, 156
as political and legal issue, 141
at societal level, 141–143, 142f.
at utility level, 141, 142–143, 142f.
See also Value of water
"Water and Related Land Resources: Principles and Standards for Planning," 101–102, 112, 193
Water and Sanitation for Health (WASH) project, 199–200
Water conservation, 73
AWWA on, 280
incentives, 82–84
metering and rate structures as incentives for, 84
Water education. *See* Environmental education
Water environment, 22–23
Water for People, 259
Water Framework Directive (European Union), 2
comparison with TWM, 61–62
comparison with US approach, 62, 63t.
leadership stipulation, 9
and normalization of water laws across nations, 230
on water pricing policies, 140
Water industry
adaptations for the future, 268
culture of, 240, 251
players who affect water resources, 29, 30t.
promoting stewardship, 272–273
stakeholders (iron triangle), 238, 251, 266, 267f.
and outreach, 191, 192f.
threats to, 265–266
trends and issues, 266–267
Water law. *See* Laws
Water management, 39
actions, 177–178, 186–187
actions that impact water resources, 24–26, 26t.
balancing utility responsibilities and corporate social responsibility, 68, 68f.
and diversions, 28, 29t., 33–34, 39

and environmental problem, 19, 22–23
and hydrologic modifications, 28, 29t., 36–38, 39
impacts on natural systems, 178–179
inadequacy of market-based approaches, 191, 203
and infrastructure, 267, 268, 273
and land management impacts, 29, 31f.
and nonpoint source discharges, 28, 29t., 35, 39
players within and outside of water industry who affect water resources, 29, 30t.
and point source discharges, 28, 29t., 34–35, 39
and risks to natural water systems, 28, 28f., 29t.
and risks to utilities and agencies, 28, 28f.
sources and impacts of actions, 30, 33f.
and sources of impact on quantity and quality, 25, 25f.
TWM as comprehensive approach to, 163–164, 165f., 186
water cycle uses, discharges, and effects, 38, 38f.
and water quality problem, 19, 21, 22t.
and water storage, 28, 29t., 31–32, 39
and water supply problem, 19, 20–21
See also Laws; Regulations; Total Water Management
Water quality, 19, 39
EPA assessment for rivers, lakes, and estuaries, 35–36, 36t.
EPA inventory of pollutants, 36, 36t.
impacts listed by source, 169, 170t.
pollution sources, 21, 22t.
Water Quality Act, 228
Water Quality 2000, 21, 169
contaminant sources identified, 22t.
on institutional issues related to nonpoint source pollution, 243–245
list of impacts by source, 169, 170t.
Water Resources Development Acts, 221–222
Water resources management
AWWA on, 280
defined, 57–58
Water resources planning, 95f., 111
art of, 101
attributes, 96, 96t.
beneficial and adverse effects, 102, 104t.
and communication systems, 108
coordination among knowledge areas and disciplines, 109
decision problems requiring, 105–106
evolution of the discipline, 96–100, 98f., 99f.
functional coordination, 109
intergovernmental coordination, 108–109
levels, 94, 95f.
negotiation and conflict resolution (consensus), 108, 111–112
and permits, licenses, and rights, 108

political model, 246–248, 247f., 251
principles and guidelines, 101–102
process (sequence), 100–101
rational vs. political, 100–101, 102f.
and shared governance, 93–94
and "silos" separating disciplines, 99–100
and stakeholder involvement, 106–107
watershed as main unit for planning, 72–75, 107–108
See also Roles and relationships; Shared governance
Water Resources Planning (M50), 98–99, 111
Water Resources Planning Act (WRPA) of 1965, 74, 97–98, 98f., 99f., 194
on multiobjective (TBL) reporting, 125, 127
Principles and Standards, 101–102, 148
Water scarcity, 165, 186
Water storage, 28, 29t., 31–32, 39
Water supply, 19, 268
demand management, 20, 39
more and less valuable needs, 20–21
Waterborne disease, 198–199, 200f., 204
Watershed management, 25, 164f., 173, 188
AWWA on, 279
as basis for planning, 72–75, 107–108
handbook, 71
micro and macro aspects, 184–185, 185f.
multipurpose basin development, 97
resource assessment, 182–186

responsibility for coordinated efforts, 74–75
Targeted Watershed program (USEPA), 186, 187t.
and TBL reporting, 129–131
as way to balance water use and ecological systems, 162–163, 164f.
See also Rivers and streams; Natural water systems
Watersheds, 172–173
defined, 172
Welfare economics, 125
Wetlands, 175–176
WFD. *See* Water Framework Directive
White, Gilbert, 59, 221, 248
Wilson, Charles, 5n.
Wolman, Abel, 74
World Bank
environmental reports, 167
on IWRM, 60–61
World Water Assessment Programme, 167

X

Y

Young, John, 6
Young, Robert, 141

Z

ABOUT THE AUTHOR

Dr. Neil S. Grigg is a professor of civil engineering and former director of the Colorado Water Resources Research Institute at Colorado State University. He is a life member of AWWA, having joined in 1970 while working as an assistant professor at the University of Denver and in his own consulting engineering firm in Denver. Neil graduated from the US Military Academy in 1961 and spent a tour of duty with the US Army Corps of Engineers. He holds an MS degree in hydraulic and structural engineering from Auburn University and a PhD in hydraulic engineering at Colorado State University (CSU). During the past 35 years he has served on the faculty at CSU, where he now heads the Department of Civil Engineering, and in North Carolina, as director of the University of North Carolina Water Resources Research Institute and for the state Department of Environment and Natural Resources. He is a registered professional engineer in Colorado, Alabama, and North Carolina and a Water Resources Diplomate of the American Society of Civil Engineers, as well as a member of numerous other professional and engineering organizations.

Grigg's professional focus over the years has been to study and write about water issues around the world. In 1988, he was appointed by the US Supreme Court as the River Master of the Pecos River, and he was a principal consultant to the Alabama Water Resources Policy Commission. In 2004 he directed a project to review water law and management in Colombia. His international water experience includes work on Egypt's master plan for Nile water use and assisting other countries in establishing graduate and research programs. He has visited a number of countries on water study tours, including the UK, Somalia, China, Vietnam, South Africa, Korea, Japan, and Indonesia. In addition, his graduate students have gone on to work in many important water management positions in a number of countries.

Grigg has published about 200 works about water and infrastructure, including several articles in the *Journal AWWA*, research for the Awwa Research Foundation, and several books, including

- *Civil Engineering Practice in the 21st Century* (with Criswell, Fontane, Siller; ASCE Press, 2001)
- *Water Resources Management: Principles, Regulations, and Cases* (McGraw-Hill, 1996)
- *Infrastructure Engineering and Management* (John Wiley, 1988)
- *Water Resources Planning* (McGraw-Hill, 1985)
- *Urban Water Infrastructure* (John Wiley, 1986)